SpringerBriefs in Applied Sciences and Technology

SpringerBriefs present concise summaries of cutting-edge research and practical applications across a wide spectrum of fields. Featuring compact volumes of 50 to 125 pages, the series covers a range of content from professional to academic.

Typical publications can be:

- A timely report of state-of-the art methods
- An introduction to or a manual for the application of mathematical or computer techniques
- A bridge between new research results, as published in journal articles
- A snapshot of a hot or emerging topic
- An in-depth case study
- A presentation of core concepts that students must understand in order to make independent contributions

SpringerBriefs are characterized by fast, global electronic dissemination, standard publishing contracts, standardized manuscript preparation and formatting guidelines, and expedited production schedules.

On the one hand, **SpringerBriefs in Applied Sciences and Technology** are devoted to the publication of fundamentals and applications within the different classical engineering disciplines as well as in interdisciplinary fields that recently emerged between these areas. On the other hand, as the boundary separating fundamental research and applied technology is more and more dissolving, this series is particularly open to trans-disciplinary topics between fundamental science and engineering.

Indexed by EI-Compendex, SCOPUS and Springerlink.

More information about this series at http://www.springer.com/series/8884

Moh Chuan Tan · Minghui Li ·
Qammer H. Abbasi · Muhammad Ali Imran

Antenna Design Challenges and Future Directions for Modern Transportation Market

Moh Chuan Tan
RFNet Technologies Pte Ltd (Singapore)
University of Glasgow
Singapore, Singapore

Minghui Li
James Watt School of Engineering
University of Glasgow
Glasgow, UK

Qammer H. Abbasi
James Watt School of Engineering
University of Glasgow
Glasgow, UK

Muhammad Ali Imran
James Watt School of Engineering
University of Glasgow
Glasgow, UK

ISSN 2191-530X ISSN 2191-5318 (electronic)
SpringerBriefs in Applied Sciences and Technology
ISBN 978-3-030-61580-2 ISBN 978-3-030-61581-9 (eBook)
https://doi.org/10.1007/978-3-030-61581-9

This Springer imprint is published by the registered company Springer Nature Switzerland AG
The registered company address is: Gewerbestrasse 11, 6330 Cham, Switzerland

Preface

The advancement in wireless technology has enabled seamless wireless communication between the data centre and end users such as public, commercial and government users. The un-coordinated wireless network infrastructures will lead to network congestion, co-channel and adjacent channel interference that degrades the wireless performance. Future proof and scalable smart wireless infrastructure are crucial to harmonise the un-coordinated wireless space involving new infrastructures and existing infrastructures. Technology upgrade always leads to multiple challenges in terms of the supporting infrastructures that are require to co-exist between the old and latest technologies. The challenges become obvious when it comes to the transportation market, where the infrastructure already exists to support the transportation-specific operation. The introduction of new wireless infrastructure has indirectly introduced more wireless devices to operate within the same premises and fight for the limited frequency spectrum. This problem gets worsen if the devices are operating in different wireless standards, where collision avoidance and transmission time synchronisation become complex and almost impossible. In additional, huge infrastructure upgrade cost is another area that requires serious attention. A smart antenna system that is field customisable with built-in beamforming features is one of the techniques aimed to minimise airspace congestion and deployment cost.

This book details the applications and deployment requirement targeted in the transportation market, where wireless infrastructures are set up for public users such as infotainment access via mobile phones and tablet through a wireless hotspot. This requires uploading huge amount of smart city-related data such as sensors, Internet of Things (IoT) devices management, etc., and government enforcements such as closed circuit television (CCTV) installed at the public area for public safety purpose. Infrastructure deployment methods and associated costs are highlighted for both the traditional method and smart infrastructure. In addition, a field configurable 360° beam steerable antenna system that can be unitised to overcome the airspace congestion issue and improve the quality of the wireless infrastructure has been presented. The book is also presenting the author's practical experience gained over the years in the wireless infrastructure deployment in the transportation market. It's a good read and reference for academic and industrial researchers and wireless system integrators as a guide to enhance and improve the wireless infrastructures in the dynamic transportation market.

This book features the wireless infrastructure deployment to support the vehicular communication and roadside infrastructure in the modern transportation market, where the wireless infrastructure is expected to sit on top and co-exist with the existing structure and details include the challenges and mitigation measures to overcome the challenges. A smart antenna structure is proposed to overcome the airspace congestion and to the end, it improves the overall wireless performance and deployment cost. With the combination of practical know-how and theoretical estimation, the book provides a good insight on how the modern smart antenna techniques that support most cutting-edge wireless technology can be adopted into the existing infrastructure that minimises the distraction to the existing system. This is in line with the global initiative to promote a smart city, where a huge number of IoT devices being wired or wireless are expected to work harmonically in the same premises. The smart antenna system structure concept presented in this book is set to deliver a future proof and highly scalable and sustainable infrastructure in the transportation market.

Singapore, Singapore — Moh Chuan Tan
Glasgow, UK — Minghui Li
Glasgow, UK — Qammer H. Abbasi
Glasgow, UK — Muhammad Ali Imran

Contents

List of Figures

Chapter 1
Introduction

The good commuting experience in the transportation sector is not only measured by the affordability of the transportation cost and the conveniences in term of the transportation networks and infrastructure framework that allow committers to reach out to every destinations within the city conveniently, ability to provide a good communication infrastructure in the transportation sector for infotainment is also a key area that is highly demanded in the modern digital world. Thanks to the advancement in the internet technology that supported by highly reliable wireless infrastructure such as Long Term Evolution (LTE), Fifth Generation Wireless (5G), 802.11ac and 802.11ax standards that have incorporated Multiple Input Multiple Output (MIMO) technologies to improve the throughput and reliability performance of the wireless infrastructure. In addition, internet-enabled devices such as smartphone, internet of thing (IoT) sensors and personal digital assistant (PDA) have further pushed the mobile broadband application to the next high. Various mobile related applications have been made available such as video streaming, internet conferencing, mobile gaming, social networking applications etc.

The proven wireless technology and user's expectation has given a strong justification for the industry to demand a highly efficient and reliable wireless infrastructure with high availability. The popularity of the wireless infrastructure, especially in the transportation sector, has led to over-congested airspace due to the increased in the wireless infrastructure. The technology advancement may turn into a disaster if the air space congestion remains unattended. Over the years, the researcher and industry partners are working hard to mitigate the air space congestion issue by various approaches, such as more coordinated wireless planning, regulatory enforcement to restrict the frequency usage and limit the radiated power, introduction of new frequency spectrum for public usage etc. The smart antenna is one of the powerful solutions that address the air congestion issue from the source itself. In the smart antenna system, both of the base station and mobile client are coordinated and only focus the radiating beam and receiving beam to each other with narrow beamwidth and null off other directions, in the transmitter, the technique is aimed to focus the

M. C. Tan et al., *Antenna Design Challenges and Future Directions for Modern Transportation Market*, SpringerBriefs in Applied Sciences and Technology,
https://doi.org/10.1007/978-3-030-61581-9_1

RF energy to the destination with minimum interference to the surrounding devices, similarly, for the receiver, it will focuses it hearing beam to the radiating source and null off all other directions to minimise the interference from the surrounding devices.

Implementation cost is another key challenge for wireless infrastructure deployment in the transportation sector due to its stringent safety requirements, high electromagnetic compatibility and immunity standard and its high availability requirement, in addition, the system integrator shall make sure minimum disruption to the existing operation while performing the infrastructure upgrade. Therefore, the equipment cost is crucial to the entire project cost, reducing the number of equipment needed without compromising the functionality of the system is another interesting area for the researchers and system integrators to continue exploring.

This book has been written for those who are working in the broadband wireless communication industry in the transportation sector that involved in deploying the smart antenna in their present or future works. Some of the industry norms highlighted in the book will be beneficial to the academic researchers, wireless infrastructure designers, infrastructure developers and equipment manufactures. This book provides a good overview in the practical usage scenarios and deployment knowhows that are related to the challenging field environment, that includes methods to ensure the robustness of the system and mitigations to overcome the challenging issues encountered in the field. This book can be a good reference as training material for the system integrators and students. For the fresh graduates, the simple and easy to understand approach includes a simple case study will create a good understanding and appreciation to the practical smart antenna deployment in the complex transportation environment.

The first task of the authors was to explore the importance of wireless infrastructure in the transportation sectors, that includes the various application in the transportation market that are making use of the wireless roadside and rail side infrastructures, application spanning from data transfer of basic vehicular information, passenger infotainment, public safety, remote monitoring and Internet of Thing (IoT) that called upon a reliable and secured wireless infrastructure. In addition, the wireless infrastructures in the transportation sector are rather complex with multiple wireless standards to serve a different kind of applications, the wireless standards include Tetra, Wi-Fi, proprietary radio, LTE/5G etc. This has created serious air space congestion if the wireless systems are operating at the same frequency spectrum with different standards. This becomes one of the challenging tasks among the system integrators and researchers to innovate a system that can co-exist harmonically with the others.

We run through the various mode of wireless communications in the transportation sector such as Vehicle to Vehicle (V2V) and Vehicle to Infrastructure (V2I) to understand how the infrastructures were deployed, their challenges and how to overcome those challenges, it was encouraging to see that the roadside infrastructure (RSU) can be converted to the sub-category called mobile RSU (mRSU) or static RSU (sRSU) to reduce the number of RSU needed while maintaining the same system performance.

The wireless cell planning is one of the topics that lead us to appreciate the advantages of smart antenna deployment compared to the conventional deployment, the smart antenna system reduces the number of equipment used and improves the wireless interference performance can be better justified. The wireless infrastructure cost proposition is explained in detail, the high equipment cost for field-installed equipment can be explained by in-depth understanding on the environmental impact and necessary field protection that is required to ensure the equipment to work reliably in the harsh transportation environment.

The requirement also specified in the respective technical specifications and standards that aimed to ensure the equipment installed in the transportation environment are meeting the minimum required standards and specifications. Therefore, a big portion of the equipment costs is contributed by additional protection or supporting equipment that is necessary to be incorporated into the system to make it robust and performs well in the harsh environment. The overall deployment cost can be summarised in the case study on the wireless infrastructure deployment for a 100 km service route. Where the overall capital investment (CAPEX) cost and operational expenditure (OPEX) cost are tabulated. The case study results again proofed that the smart antenna deployment will not only improve the interference performance but also reduce the system deployment cost drastically for both CAPEX and OPEX.

The key challenges and how the smart antenna system can help to mitigate those challenges are highlighted, in particular to the transportation sector, we have identified the operation scenarios that allows the smart antenna system to leverage on with some pre-configurable features that can further improve the cost and wireless interference performance compared to just by smart antenna itself.

Chapter 3 covers the research on the existing state of the art antenna systems, we have widened the study into 3 areas, the 360° coverage smart antenna system, the antenna array and the antenna element that build up the array. Finally, we decided on our smart antenna structure "A modern pre-configurable smart antenna system" which allow pre-configuration on both gain and coverage sector and perform electronic beam steering up to 360° in the azimuth plane.

Chapters 4 to 6 cover the necessary steps to design the antenna system, start with the single element design, simulation, and optimisation, followed by integrating the elements into 3 types of array 1×4, 2×4 and 4×4 with low, medium and high gain. The size of the antenna ground plane that determines the overall size of the antenna and the elements separations are simulated to achieve the optimum performance, the trade-off between the gain, bandwidth and sidelobe performance are highlighted.

The prototype of the antenna arrays are fabricated using commercially available printed circuit board (PCB) fabrication process followed by the soldering process to construct the array that consists of 2 pieces of PCBs, and finally, both of the single elements and arrays are experimentally evaluated on their conducted and radiated performance. The measurement setup and topology are explained in detail with supporting photos. The performance of the 360° pre-configurable antenna system was evaluated and tabulate in a table to compare with the state of the art 360° beamforming antenna systems.

Chapter 7 provides a moving forward overview of the smart antenna revolution in the modern transportation sector and conclude the book. In the IoT world, the environment is expected to be packed with IoT related sensors and communication devices including short-range and long-range communications systems. The common questions posted by the stakeholder or infrastructure planner are how much we can leverage on existing infrastructure, sensors and make use of it to minimise the introduction of the new equipment and lower the overall project cost, we are trying to answer this question with our proposed future direction on the modern beamforming antenna system.

By taking the advantage of the pre-installed and fixed roadside or rail side infrastructure, in addition to the fixed moving route of the public transportation such as buses and trains, we are anticipating a futuristic smart antenna system that makes use of this operation advantage combining with the IoT sensors such as global positioning system (GPS), accelerometer, gyroscope, odometer etc. The sensors data are combined via a sensor fusion module together with the information of the known locations of the roadside infrastructures, we can generate the useful location-based information that helps the roadside base station and mobile terminal to coordinate and determine the accurate beamforming angle within them while the mobile clients are on the move. With the pre-defined routes, known locations, speed and heading, predictive roaming and antenna beamforming angle between base stations and mobile clients can be easily determined, by doing so, we can reduce the highly complex field-programmable grid array (FPGA) cost that is usually used in the conventional beamforming smart antenna system. This is expected to reduce the smart antenna cost further while proving more accurate beamforming capability.

Chapter 2
Wireless Infrastructure in the Transportation Market and the Challenges

2.1 Wireless Infrastructure in the Transportation Market

As the wireless technology emerges, sending the big amount of data from mobile terminals over the wireless infrastructure to the backend server becomes possible, this has created a lot of opportunity in digital data mining, migration from analog system to digital system and automates partial of the operations and control via wireless infrastructures, this is in line with the local government initiative to promote the smart, connected and sustainable smart city. This has improved the operations of the local industries such as transportation, factory automation, autonomous vehicles, city surveillance etc. Among all, the transportation section has benefited most due to the nature that the automotive vehicles are physically isolated from the ground system and the only mode of communication is via wireless. A book on Railway related Broadband Wireless Communications [1] by Railenium, the French Innovation, Research and Technology (IRT) has demonstrated the equipment and infrastructures that were installed to cover the in-vehicle internet access and other related applications that aimed to bridge between the industry and academic research, the coverage including the railway wireless standards requirement as well as operational applications and passenger applications.

Another book [2] has consolidated the works related to the vehicular communication and networks, the topics covered by this book include Vehicle to Vehicle (V2V) and Vehicle to Infrastructure (V2I) communication, medium access control (MAC) layer in vehicular communications, vehicular communications protocols such as opportunistic routing, dynamic spectrum access, security and privacy, the connected vehicle in intelligent transportation system (ITS) and some simulation and modelling work in vehicular networks and location-based applications, this provides the insight with join opinions from different authors who are expert in their research field in particular to the vehicular networking.

In the transportation market, wireless infrastructures are deployed for various type of applications, the most common applications are highlighted below.

M. C. Tan et al., *Antenna Design Challenges and Future Directions for Modern Transportation Market*, SpringerBriefs in Applied Sciences and Technology,
https://doi.org/10.1007/978-3-030-61581-9_2

Smart City, where various type of sensors were deployed at the roadside and vehicles to constantly upload device health data to the backend for device health monitoring, control and big data analysis, for example, the smart street light, where the ambient sensor is deployed on the street lamp to monitor the environmental condition such as temperature, humidity, dimmer level, etc., all the smart lamp poles are wirelessly linked and the data collected will be sent to the backend wirelessly for analytic. For smart car park application, the cars are equipped with wireless sensors that communicate with the car park infrastructures for parking charges and parking management and monitoring.

Power meters and waters meter turned smart with built-in wireless, all the smart meters in a city will be wirelessly connected to the back end management server, where remote monitoring and meter reading can be done easily that is essential to reduce the human cost. Another example is for disaster management system, where multi-functional sensor such as seismic sensors and flood sensors can be installed at some prominent area to have early detection and warning on earthquake, flood and typhoon that can save millions of humans life if the disaster can be detected earlier and warning can be disseminated on time. All the sensors can be controlled and managed wirelessly via a central device management system located remotely.

Closed-circuit television (CCTV), as part of the public security initiative, local government and enforcer has installed public CCTV along the public roadside and within the mobile vehicles to monitor the crowd condition and ensure the safety of the residents and passengers. In the past, the CCTV footage was stored in the local storage and re-cycle itself due to the limited storage, with the advancement of wireless technology and cloud storage, the CCTV footage will be sent to the backend server for storage and further video analytic process, artificial intelligent (AI) and machine learning (MI) can be incorporated at the backend with powerful machines such as high-end graphic processing unit (GPU), FPGA for video analytic, MI and inferencing, such equipment is not possible to be installed at the edge or mobile terminal due to their high cost and power consumption.

Passenger Information, most of the public transports, stations, terminals and interchanges had equipped with passenger information display system (PIDS), it can be in the form of audio and visual display such as light-emitting diode (LED) display or graphic liquid crystal display (LCD), passenger information such as current stop, next stop, interchange information, time of travel, emergency announcement and intercom are disseminated to provide commuters with the timely update and create a pleasant and comfortable journey. The modern transportation system also come with large signages were the stations' information and nearby attractions can be pushed to the signages that are located at the prominent area. Due to the locations of the PIDSs are scattered around the city, hence wireless communication is the most convenient communication network to connect them with the back end.

Passenger Infotainments, the advancement of wireless technology has also improved the lifestyle of the commuters, the mobile terminals installed in the automotive vehicle such as public buses and trains are wirelessly linked to the backend server, the mobile clients are expected to roam seamlessly across the wireless infrastructures installed along the roadside or rail side. Many types of public transports are equipped with signage for advertisement and posting of real-time announcement or commuter related information. Wireless hotspots are available within the public buses and trains to provide entertainment during the bus or train journey such as internet serving, video on demand (VoD), voice over IP (VoIP), live streaming, video/television conferencing etc. This has indirectly jacked up the usage expectation where a reliable infrastructure is expected to overcome issues such as air space congestion and un-coordinated infrastructures deployed by different service providers.

Operations and Services, this is related to the command and control of the vehicles, including schedule planner, destination routing, signalling, obstacle detection, driving assistance, collision warning system (CWS), fare management system, driver management systems, real-time information exchange between the mobile vehicles and backend allows remote monitoring and automates part of the operations remotely. For preventive and corrective maintenance, the central maintenance system can remotely record and monitors the sensors installed onboard, such sensors include, tyre pressure sensors, flood sensors, CCTV footage that monitors the entire sector of the rail structure for early detection of crack or deformation, the sensors data collected from the road or rail side can be analysed as part of the predictive and corrective maintenance exercise that improve the system reliability.

A flat functional map of the various communications component in the modern transportation sector is presented in Fig. 2.1. In the practical scenario, the automotive vehicles are on the move and it requires a seamless communication with the backend that provides the communication link between the vehicle to the backend Operation and Command Center (OCC), uploads vehicular information, and upload data collected from the journey as well as provide infotainment and internet connectivity for the commuters during their journey, all the above require a stable, reliable and robust wireless infrastructure. The hard fact is that the communication media adopted in the transportation environment usually comes with different wireless systems that work on different standards and operating frequencies that make the co-existence task challenging, the network or wireless engineers need to design the system very carefully to ensure the co-existence between different communication equipment

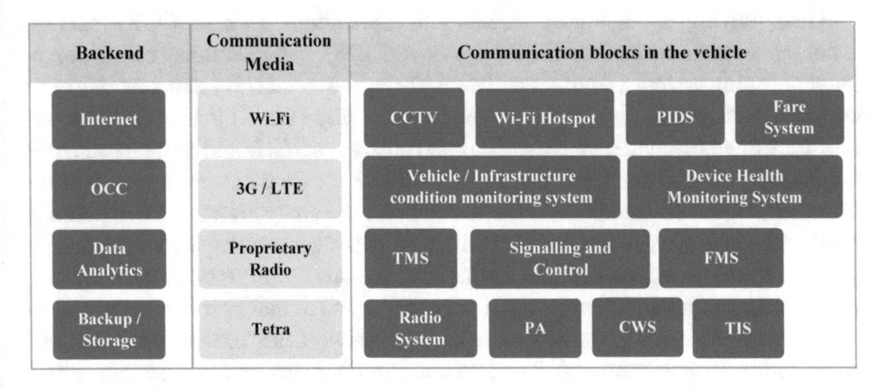

Fig. 2.1 Typical communication components in a public vehicle

with multiple wireless standards to work harmonically within the same premises in the transportation environment.

In the transportation sector, Trans-European Trunked Radio System (TETRA) radio is widely used as radio system for data and voice communication, voice communication includes the communication between the field operation personnel and the control center, public address (PA) announcement etc., the TETRA radio also used as a data communication link to transfer a small amount of data over the network such as short emergency messaging, device health monitoring etc. The TETRA system operates in the very high frequency (VHF) band around 150 MHz and ultra high frequency (UHF) band from 300 MHz to 1 GHz, the frequency selection highly depends on the local authority who allocates the frequency spectrum in the respective country. The communication distance of the TETRA system depends on the allowable effective isotropic radiated power (EIRP) allowed by the local government, however, due to the radio propagation nature of the VHF and UHF frequency, the signal can be penetrated very far even for operation scenario where physical line of sight coverage is not possible such as the congested transportation environment, tunnel and within the buildings, this has made the TETRA radio a popular choice for mission-critical voice and data communication in the transportation sector.

For other mission-critical communication such as train management system (TMS), train signalling and control, interfacing with train information system (TIS), fleet management system (FMS), CWS etc. that require higher data throughput, low latency, and secured connections, The proprietary radio communication system that may operate in different frequencies such as 2.4 GHz etc. will be used, however, the choice is highly depending on the transport operators itself. Such proprietary radio has built-in necessary information technology (IT) security profiles and collision avoidance that makes it reliable to handle the mission-critical communications. Usually, such a system comes as a standard built-in system provided by the vehicle manufacturer such as rolling stock and bus manufacturer as a standard package. Due to most of the usages are for mission-critical applications, the proprietary system was developed a long time back and its reliability and safety have been fully tested, verified, and proven over the years. In general, such a system will not be easily replaced.

When moving up to higher throughput applications such as CCTV surveillance, passenger information display systems (PIDS), vehicle/infrastructure condition monitoring systems, device health monitoring system, fare systems etc. wireless networking standards such as 802.11 wireless fidelity (Wi-Fi) that operated in the Industrial, Scientific and Medical (ISM) band on 2.4 and 5 GHz and cellular standards such as LTE and 5G can be deployed. The 802.11 standards are preferred than the cellular due to the licensed free spectrum and commercially affordable compared to cellular where the transport operators will need to pay for monthly service subscription fees to the telecom operators throughout the operation years. Over the years, we are seeing more adoption towards the cellular usage thanks to the reduction of subscriptions cost due to massive competition amongst the telecom operators and the reduction in equipment cost. The data collected from the train will be transmitted in real-time, delayed or opportunist transmission to the back end or Operation and

Command Center (OCC) where remote monitoring, command and control or storage and further data analytic take place. In the IoT world, for instance, the data collected from vehicle/infrastructure monitoring system and TIS can be further processed using data analytic, machine learning (ML) and artificial intelligent (AI) to generate visualisation in term of the graphic user interface (GUI) or interactive dashboard to provide the stakeholder with an overview of the overall system performance.

With the help to technology, the big data can be further translated into useful data to determine the likelihood of components failure that may lead to the system failure which can be further translated into predictive and preventive maintenance to further enhance and improve the reliability of the system. With the powerful processor at the backend, the CCTV footage transmitted to the backend can be further processed using highly complex video analytic (VA) technology. The VA with the help of ML and AI allows the system to provide alerts when it identifies any suspicious objects or persons, public bullying etc. this can serve as part of crime prevention activities in the transport industries.

The smart gadgets had emerged rapidly into our daily life, smartphones or tablets with internet connectivity becomes part of our accessories when we are in the moves. Lifestyle has been improved and the user expectation in the internet connectivity has risen, the PIDS provides real-time information about the journey, travelling time, current location and destination as well as providing geographical highlight especially for the tourists. The in-vehicle wireless hot spot becomes popular with the introduction of smart devices, while in the moves, the commuters are expected to keep in touch with their friends through the social platform, perform works or entertainment that requires seamless connectivity to the internet. The Wi-Fi hot spot usually operates in 2.4 GHz or 5 GHz ISM band, which is the built-in feature in the modern smart devices. The internet connectivity of the mobile vehicle is obtained through a backhaul wireless connection to the backend either via another Wi-Fi infrastructure or cellular such as 4G. In recent days, new infrastructure deployment has considering Wi-Fi as roadside or trackside infrastructure to provide seamless data connectivity when the vehicle is in the moves.

2.2 Mode of Communications

The wireless communications in the transportation sector can be divided into 2 major areas, V2I and V2V. In the V2I infrastructure, the roadside unit (RSU) is installed at the roadside or nearby building to provide communication between the vehicular networks and the moving vehicles it's convenience to use existing cellular network or Wi-Fi technology to handle the necessary data throughput. In the case of V2V communications, the preferred technology can be LTE device to device standards or Dedicated Short Range Communication (DSRC) technology, and it has gradually set a foundation for the future autonomous vehicle. In this section, we have included some interesting works that were presented in the academic world to demonstrate and enhance the communication performance in the V2I and V2V communications network.

In [3], the authors had demonstrated the comprehensive and comparative study in a city environment cover the communication hardware and software details as well as system integration models for two important vehicle communication types, V2V and V2I communications. The software-defined network (SDN) concept that enables efficient data services in vehicular networks was defined. The networking scheme presented includes DSRC for the vehicle to vehicle communications, In-cabin Wi-Fi for hot spot applications and vehicle light communications. The book also covers the type of antennas or V2V and V2I communication and the physical layer according to the IEEE 802.11p standard, the IEEE 802.11p V2V and V2I communication technology is an amendment to the IEEE 802.11 standard that added Wireless Access in Vehicular Environments (WAVE) in Wi-Fi-based to support ITS applications.

The V2I and V2V communications based on Heterogeneous Vehicular NETwork (HetVNET) frameworks have been illustrated in [4] for the requirement in safety as well as non-safety services in the ITS, the V2I and V2V communication were evaluated based on LTE cellular and DSRC technology which provides real-time and efficient information exchange among vehicles. By evaluating the various protocols on medium access control and networks layers, the authors have concluded that the LTE and DSRC are the best solutions to support the HetVNET networks for V2I and V2V communication respectively, where the cellular network provides wide coverage for the automotive users and the DSRC communication provides the reliable connection between vehicle to vehicle with low latency which is crucial for safety–critical communications.

In practical vehicular communication, the cellular networks provide wide coverage range to the vehicular, however, it has a limited bandwidth when the number of vehicles that need to be served by a single base station increases, the bandwidth offload via roadside Wi-Fi infrastructure was proposed in [5] where the vehicles will aggressively cache the data when it passes the base station for offline browsing, the experiment was done in the highway and the results revealed that both unicast and broadcast-based systems have it communication throughput degraded when the density of the vehicle increases, the authors have proposed a method to incorporate device-to-device data scavenging that allows the vehicle to share information within them when the vehicles are passing the base stations, the method has successfully improved the throughput and reliability of the connections.

One of the main functions of the roadside infrastructure is to upload the CCTV data or provide real-time viewing of the CCTV footage in the vehicles at the backend 802.11p in the vehicular environment. H.264 also referring to Advanced Video Coding (AV) is a most popular standard for coding of High Definition (HD) digital video, it compresses the video to approximately half of the space of MPEG-2 (the DVD standard) and maintaining the same high-quality video, H.264 is a very common coding technique used in CCTV surveillance system. In [6], the H.264/AVC codec extension has been proposed, the proposal was based on the 3-D discrete wavelet transform comes with a rate control algorithms and low-complexity unequal packet loss protection to achieve scalable video coding, good video visual quality was achieved with the V2I communication distance within 600 m.

The roadside infrastructure is also referring to the RSU, as the cost of the roadside deployment is very high due to the stringent requirement of the equipment in order to sustain in the harsh outdoor environment and fieldwork such as cabling and equipment mounting structure. The study was carried out to reduce the deployment cost in [7] by introducing the concept of sub-category RSU called mobile RSU (mRSU) and static RSU (sRSU). By taking the advantage of the buses that have a fix travelling route, bigger volume to house the RSU, the experiment and analysis results on the replacement ratio has been conducted based on the contact time, intercontact time and throughput. The results revealed that the mRSUs can replace sRSUs with certain replacement ratio without degrading the system throughput. The results can be used as a guideline to determine the utilisation ratio of mRSU and sRSU that achieve the optimum performance and cost trade-offs.

2.3 Wireless Cell Planning

The wireless cell panning is commonly used in the wireless infrastructure deployment planning to determine the number of base stations is needed to provide optimum coverage to the service area, performance criteria such as coverage range of each base station, reuse of frequency or channel etc. will be considered during the wireless cell planning. In the conventional wireless cell deployment, to ensure minimum interference from adjacent cells and maximise the usage of available frequency channels, a set of non-overlapping frequencies are used in the cluster, each cluster consists of multiple hexagonal cells, the cluster patterns are reused to create the entire coverage area. A low-cost hybrid beamforming antenna structure has been presented in [8], the paper highlighted the cell planning requirement for conventional and smart antenna system and the benefits with the smart antenna are highlighted along with the low-cost hybrid beamforming architecture. The cells planning scheme for 3 non-overlapping frequencies and 7 non-overlapping frequencies are shown in Fig. 2.2, different frequencies are represented by different colours. The cell deployment scheme can be chosen depending on the type of communication system, for instance, 7 non-overlapping cell planning scheme can be deployed on 802.11a/ac system that operates on 24 non-overlapping channels, while 802,11bgn with only 3 non-overlapping channels shall adopt 3 non-overlapping cell planning scheme. The 7 non-overlapping cell planning scheme has the advantage over the 3 non-overlapping cell planning scheme in term of co-channel interference due to the frequency reuse is 3 cells away compare to only 2 cells away for the 3 non-overlapping cell planning scheme.

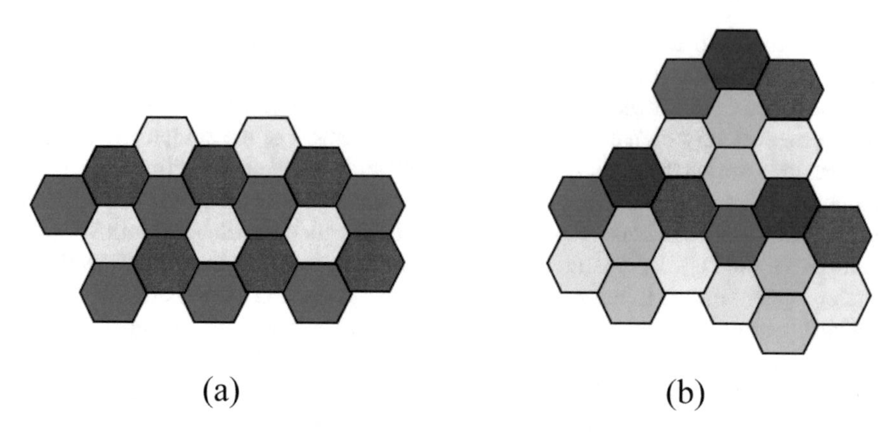

Fig. 2.2 Cell planning. **a** 3 non-overlapping channels. **b** 7 non-overlapping channels

In the practical deployment, the cell size was calculated using Free Space Lost formula in [9] and include the environmental loss factor.

$$\mathrm{FSL} = 20\log\left(\frac{4\pi f d}{c}\right) \tag{2.1}$$

where:

f: frequency of interest.
d: distance between the base station and client.
c: speed of light (3×10^8 m/s).

In most of the practical wireless deployment, especially in the transportation environment, the base station is required to communicate with multiple mobile or stationary clients within the service area. Figure 2.3 shows the conventional method

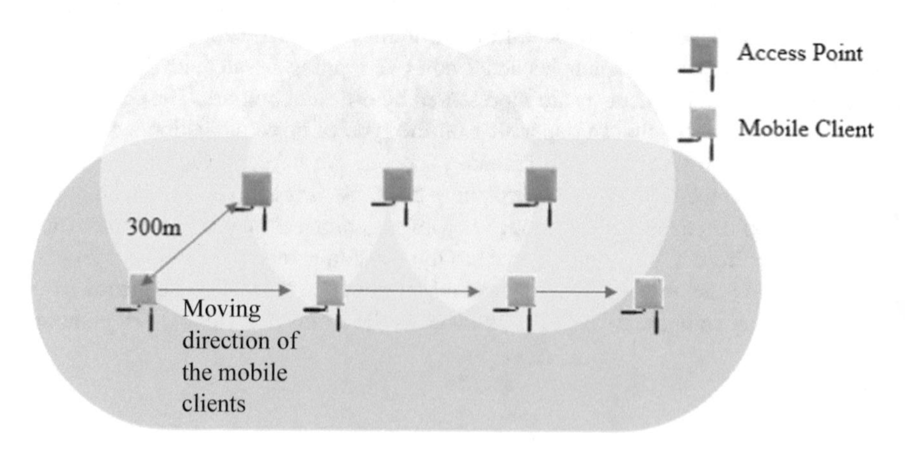

Fig. 2.3 Conventional Point to Multipoint deployment

to set up the base station at roadside or tunnel to serve the moving vehicle or train, omni antenna is used for 360° coverage. Yellow cells represent the access points (AP) coverage and orange cell represents the coverage for the mobile clients. There will be overlapping between the APs coverage zone, this is to ensure it can provide maximum coverage to the entire route that the mobile clients are expected to travel.

For traditional AP deployment, the APs and the mobile terminals are both transmitting at 17 dBm power with 10 dBi antenna gain, the calculated cell size is approximately 300 m. Imagine the mobile client is moving from the left side of the coverage zone to the right, and it supposed to roam across the access points installed along the roadside. The potential issues anticipated from the existing deployment are:

i. Small communication radius leads to more AP is needed to cover the entire service area. For example, a 10 km road requires around 34 AP.
ii. Mutual RF interference between the adjacent cells if they are operating on the same frequency.
iii. Higher installation cost and more installation materials.
iv. Higher CAPEX and OPEX contributed by more equipment.

To overcome the current limitations, a smart antenna system is proposed, which allows the AP to steer the RF beam towards the mobile client and interference nulling by making use of beamforming technique, in addition, the concentration of transmitter and receiver radiating beam has concentrated the radiating power to a narrow beam which translated into higher antenna gain, the concentration of radiating beam also provides a good interference immunity as the transmitter and receiver only radiate and listen to the narrow intended beam. Adaptive beamforming can be incorporated to dynamically steer the beam on-the-fly by tracking the client direction using the direction of arrival technique. With this technique, when the clients are roaming within the service area, both the radiating beam of the access point and mobile client will steer and adjust to each other. Figure 2.4 exhibits the radiation beam of the access points and client dynamically steer to follow the direction of the access points and clients.

The benefits gained from the smart antenna are:

i. Longer coverage distance due to the higher gain antenna, and thus fewer access points are needed. For example, a 10 km road requires 5 access points.

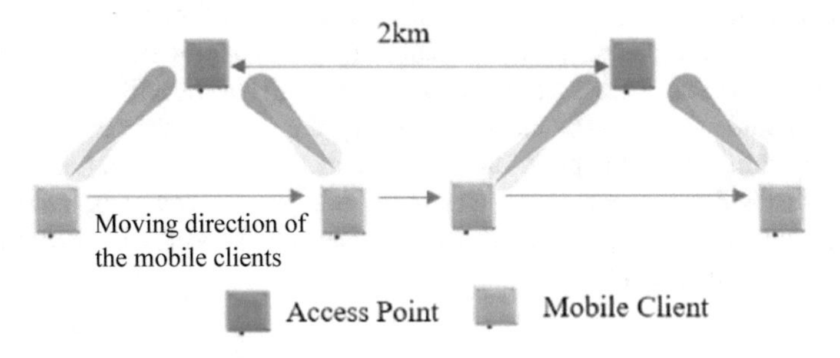

Fig. 2.4 Deployment with the smart antenna

Table 2.1 Coverage estimation for access points serving moving clients

Type of antenna	TX power (dBm)	TX antenna gain (dBi)	Distance (m)	RX antenna gain (dBi)	Calculated RSSI (dBm)	Target RSSI (dBm)	Margin (dB)
Conventional (omni antenna)	17	10	300	10	−59.8	−70	10.2
Access points with smart antenna and mobile clients with omni antenna	17	18	800	10	−60.3	−70	9.7
Smart antenna on both access points and mobile clients	17	18	2000	18	−60.3	−70	9.7

Note TX Transmitter, *RX* Receiver, *RSSI* Receiver signal strength indicator

ii. Adjacent cell interference reduces due to narrow beamwidth.
iii. Low deployment and maintenance cost.
iv. Energy saving in equipment usage.

Table 2.1 estimates the useable link distance between the access point for conventional deployment and smart antenna deployment serving the wireless clients in the transportation environment. 10 dB margin was included to act as the buffer for possible attenuation caused by environmental factors such as building, tree, moisture etc. The calculated cell size for the conventional omni antenna deployment is approximately 300 m, whereas the coverage increases to 800 m if the access points are equipped with smart antennas while the mobile clients are installed with omni directional antenna, the range can be further improved to 2 km if both the access points and clients are equipped with the smart antenna. This simulation is based on the theoretical line of sight simulation, in real life, the signal propagation loss due to non-line of sight scenario must be considered.

The benefit of the smart antenna in the transportation sector is obvious. The improvement in the coverage distance with smart antenna technique implies lower CAPEX due to fewer access points is needed and lower deployment cost because of minimum equipment and less wiring is needed, this will also benefit the OPEX where less equipment is needed to be taken care during the whole maintenance regime. In the next section, we will explore the components that constitute to the cost of the infrastructure's deployment, that includes the equipment need to be environmentally protected as well as the requirement that specified in particular environmental specifications and standards.

2.4 Wireless Infrastructure Cost Proposition

The deployment of wireless infrastructure in the transportation sectors can be divided into two schemes, new infrastructure, and upgrade on existing infrastructure. Introduction of new infrastructure is straight forward where new infrastructure such as communication equipment and new cabling and automotive stocks will be introduced, the project owner usually will do the pre-assessment on the overall wireless deployment plan taking care of the various type of applications and co-existence between multiple wireless standards, wireless cell planning can be coordinated easily within the same contract or tender. More challenging when coming to an infrastructure upgrade, infrastructure upgrade or commonly called as technology refresh that comes with the introduction of the latest technology into the existing systems while preserving the existing wireless system that is currently supporting the operations. In this context, 802.11ac/ax, wireless gigabits (WiGig) or fifth-generation (5G) wireless technology is introduced into the existing infrastructure that may still operate in older technology such as third-generation (3G) wireless technology, 802.11a/b/g or due to the old infrastructure was unable to support the increasing demand in broadband wireless communication which was driven by the advancement in wireless technology.

The new technology being introduced may operate in the same frequency band as the current wireless systems that already in place to serve the operational needs of the existing infrastructure. The introduction of new wireless infrastructures has indirectly introduced more wireless devices to operate within the same premises and fight for the limited frequency spectrum, the problem gets worsen if the devices are sharing the same frequency spectrum but operating in different wireless standards, where collision avoidance and transmission time synchronisation become complex and almost impossible. Therefore, the system integrator (SI) is required to perform the wireless cell planning very carefully with the existing stake owner taking into consideration the limited frequency spectrum being allocated by the Infocomm authority.

The associated cost for both deployment scheme can be further derived by considering their CAPEX and OPEX, in most of the township or government project, the CAPEX is the main considering point while keeping the OPEX as low as possible. When comes to the infrastructure upgrade, the SI will need to consider deploying new cables and equipment on top of the existing structure, this will require additional efforts and cost to assess the interference that may potentially be introduced to the existing system due to the new equipment as well as protecting the current structure or cable that was deployed previously to support the existing equipment.

2.4.1 What Constitutes to the Cost of the Communication Infrastructure?

In the transportation environment, the infrastructure cost is not simply the cost of the equipment itself, the cost is distributed over to every single step that required to ensure the infrastructure can sustain the harsh environment, minimise the interference to other equipment and importantly it does not produce any hazards to the public, additional supporting equipment is needed to protect the system which constitutes a major portion of the cost. The next few paragraphs highlight some major requirements, specifications or standards that are mandatory for every equipment to fulfil before it can be installed at the field.

Environmental Protection, taking a closer look at the transportation environment, in general, the equipment is mainly installed outdoor with environmental protection. The outdoor environment is subjected to the effect of weather change such as temperature, humidity, rain, sunshine, snow etc. Being part of an equatorial region such as Singapore, thunderstorm, high humidity, and severe rainfall are affecting the reliability of the equipment. Northern area such as the United States, Europe, China, and Japan may experience wider temperature change across the year with snow during the winter and temperature more than 40 °C during summer. Therefore, it's costly for such kind of equipment that requires special treatment electronically or mechanically to fulfil the environmental protection or international protection (IP) as specified in the international standards International Electrotechnical Committee (IEC) 60529, the IP rating specified the degree of intrusion protection against the solid objects, accidental contact, dust, and water into the electronic device housed inside the enclosures. In [10], it was highlighted that the Trade Association for UK or Railway Industry Association (RIA) that suppliers the equipment and services to the railway industry has published the environmental standards IEC 61373 to deal with the environmental protection required for the equipment in the railway environment. As of today, the IEC 61373 standards is widely adopted by the equipment manufacturers across the world.

Electromagnetic Compatibility and Electromagnetic Interference (EMC/EMI), the transportation environment is also experience severe EMC interference, especially in the rail environment, where the high traction voltage that powering the rolling stocks will generate the huge amount of EMC that may potentially interfere and degrades the performance of the nearby electronic equipment. Another source of EMC surge is from lightning, were a huge amount of electromagnetic surge can be induced or coupled via ground or cable into the nearby equipment and potentially damages the electronic inside the equipment. Lightning arrestor or EMI suppressor can be installed to isolate the equipment from the vulnerable parts of the system. On the other hand, every electronic equipment will generate unwanted electromagnetic wave when in operation, the equipment that installed in the transportation environment must be designed with EMC treatment to control and restrict the unwanted EMC radiation below the allowable limit, this is crucial in order not to interfere with other mission-critical equipment that forms part of the transportation system to ensure safe operation of the vehicles.

The EN 50121 standards [10] addresses and defines the EMC/EMI requirement in the railway application environment, the family standards split into different parts of standards that deal with the different area of EMC/EMI requirement, for instance, EN 50121-2: Electromagnetic Emission to the outside environment, EN 50121-3-1 defines the Electromagnetic Emissions from Rolling Stock, EN 50121-3-2 defines the Electromagnetic Environment for on Board Equipment, EN 50121-4 defines the Electromagnetic Environment for Signalling and Telecommunications Equipment, EN 50121-5 defines the Electromagnetic Environment for Fixed Installations Equipment, and many more.

The requirements of environmental protection, EMC/EMI and surge immunity described above will be translated into the total project cost. For example, a smart antenna that intended to be deployed in such environment will need to have an IP rated enclosure, built-in lightning arrestor, EMC/EMI suppressor that is necessary to isolate or suppress any EMC/EMI issue in the field, the outdoor equipment will be costly due to the usage of industry-grade components to guarantee its performance and survival in the harsh environment. Usually, the equipment manufacturer will need to certify the product in the third-party accredited test lab before it can be deployed and operates in the transportation environment. The additional environment protection design and certification is expected to take up 50% of the overall equipment cost. Therefore, it is obvious to say that the number of equipment needed in a transportation system is key to determine the overall system cost.

2.4.2 Project Cost Estimation—A Case Study

A report on smart antenna deployment was presented by Vector Fields Limited to Ofcom in 2004 [11] far before the smart antenna technology was widely used in the commercial word. The project was initiated to investigate the adaption of the adaptive smart antenna in the broadband wireless access (BWA) and wireless area network (WLAN) or Wi-Fi, the evaluation was carried out using the 802.11a hardware. The study result shows that there was a significant benefit in smart antenna technology, the benefits are explained in 3 areas, (i) the increase in the communication cell range of 40–70%, (ii) reduction in spectrum utilization by 60–70% and (iii) network cost reduction by up to 50% that translates to a total business cost reduction of up to one third. This has provided positive motivation to the smart antenna industry significantly.

To better explain the cost propagation for the infrastructure deployment in the transportation market, a practical case study is presented here by comparing the infrastructure deployment using traditional and smart antenna techniques explained earlier in Figs. 2.3 and 2.4, the case study will also include all the necessary consideration that makes up the systems cost such as cabling, equipment protections, test and commissioning, as well as maintenance cost. The case study was carried out to demonstrate the general wireless infrastructure deployment in Singapore transportation sector targeting the wireless infrastructure for public buses and railway system. The wireless system proposed in this study is 802.11ac wireless LAN operating in 5 GHz band. The case study is to serve as a guideline for general deployment using 802.11ac access point, the cost may vary depending on the requirement, type, and the size of the contract. The maintenance cost tabulated below also highly depends on the maintenance regime agreed between the stakeholder and the contractor. The breakdown of the project implementation cost is presented in Table 2.2.

The regulatory certification cost is a one-time cost that is applied to all the products to be deployed in the field, the equipment cost includes the main access point product, antenna, EMC protection device, lightning arrestor, and environment protection enclosure to house all the supporting devices. Installation cost is derived by taking into consideration the field-installed devices will need to draw power and data through cable from the intermediate device box (IDF) that includes the labour cost to install the devices and its cabling, the data is connected to the back end via ethernet connection, it's necessary to use armour grade type of cable or minimum Category-5 (Cat-5) shielded S/FTP with protective trunking. The installation cost includes the mounting brackets and poles that erect the antenna to a certain height for optimum RF performance. Test and commissioning is a process designed to test and make sure the configurations of the equipment are set correctly and the performance of the system is up to its expectation.

The setup cost, equipment cost, and installation cost are categorized under the CAPEX cost. The OPEX is mainly referring to running cost incurred over the operation years, the OPEX cost includes corrective and preventive maintenance and extended warranty, the warranty of the infrastructure equipment in the transportation market usually called for a minimum of 7 years period, however, most of the electronic equipment are providing 1–3 years of limited warranty, therefore, the extended warranty cost shall be provided to give provision of warranty up to 7 years or beyond.

Finally, we can see that the CAPEX investment can achieve approximately 39% cost saving for hybrid smart antenna system while cost saving of 54% for full smart antenna deployment, more importantly, the smart antenna system has drastically improved the OPEX by approximately 63% for hybrid smart antenna system and nearly 83% for the full smart antenna system. The cost benefit brought by the smart antenna technology can potentially create a huge opportunity for the business owners where the return of investment (ROI) can be easily justified.

2.5 What Are the Challenges?

Now we are clear that the main challenges for the infrastructure deployment in the transportation sector are the air space congestion and deployment cost, below are the summary of the challenges and mitigation objective that we are trying to achieve with the smart antenna system.

Table 2.2 Deployment cost for the roadside infrastructure covering 10 km distance (in US$)

	Items	Cost per unit	[a]Scenario 1 (35 APs)	[b]Scenario 2 (13 APs)	[c]Scenario 3 (6 APs)
Setup cost					
1	Regulatory certification	$ 50,000	$ 50,000	$ 50,000	$ 50,000
Equipment cost					
2	Industrial 802.11ac wireless LAN	$ 750	$ 25,250	$ 9,750	$ 4,500
3	Omni antenna	$100	$ 3,500	–	–
4	Smart antenna frontend	$ 400	–	$ 5,200	$ 2,400
5	EMC filter	$ 120	$ 4,200	$ 1,560	$ 720
6	Lightning arrestor	$ 50	$ 1,750	$ 650	$ 300
7	Environmental proof enclosure	$ 300	$ 10,500	$ 3,900	$ 1,800
Installation cost					
8	Cat-6 cable (material and installation cost per 100 m)	$ 1,000	$ 35,000	$ 13,000	$ 6,000
9	Structure to mount the equipment	$ 500	$ 17,500	$ 6,500	$ 3,000
10	Test and commissioning	$ 100	$ 3,500	$ 1,300	$ 600
Total CAPEX			**$ 151,200**	**$ 91,860**	**$ 69,320**
Cost benefit on CAPEX			–	**39%**	**54%**
Maintenance					
11	Corrective maintenance for each wireless access point (yearly)	$ 200	$ 70,000	$ 26,000	$ 12,000
12	Preventive maintenance for each wireless access point (yearly)	$ 100	$ 35,000	$ 13,000	$ 6,000
13	Extended warranty (yearly)	$ 75	$ 2,625	$ 975	$ 450
Total OPEX			**$ 107,625**	**$ 39,975**	**$ 18,450**

(continued)

Table 2.2 (continued)

	Items	Cost per unit	[a]Scenario 1 (35 APs)	[b]Scenario 2 (13 APs)	[c]Scenario 3 (6 APs)
Cost benefit on OPEX			–	**63%**	**83%**

Source Training material that consists of general costing guideline for the transportation-related project in RFNet, Singapore
[a]Scenario 1: Conventional deployment with omni directional antenna
[b]Scenario 2: Hybrid deployment with access points equipped with the smart antenna and mobile terminals deployed with omni directional antenna
[c]Scenario 3: Deployment with smart antenna equipped on both access points and mobile terminals

i. **High CAPEX**, reduce the number of field equipment will significantly contribute to the reduction of the CAPEX cost, this can be done by introducing the beamforming technique into the antenna systems, further cost reduction on the smart antenna system that allows pre-configuration to allow only necessary sub-array to be installed.
ii. **Wireless Interference**, Reduce the Mutual interference between adjacent cells if they are operating on the same frequency, this can be achieved by beam steering where the radiating beam of the transmitter and receiver antenna are always pointing to each other with narrow beam and nulls all other direction.
iii. **High OPEX**, as discussed earlier, the smart antenna technique has significantly reduced the number of equipment needed to be deployed, this will lead to a reduction on the OPEX cost, operating cost such as preventive and corrective maintenance can be controlled if there is less number of field equipment need to be maintained.

2.6 How the Smart Antenna Structure Helps

There is always a fixed constraint on the wireless link between mobile terminals in the moving vehicles and the base stations installed along the roadside, this is mainly due to the mobility of the mobile terminals, the antenna beam for the roadside infrastructures and the mobile terminals are not always fixed. As described in Table 2.1, the traditional wireless deployment was implemented by an omni-directional antenna that provides 360° coverage for the access point and the mobile client, this method has limited the coverage radius of the base station and the omni antenna implies the base station can transmit and receive the signal to and from all the direction, the base station can be easily interfered and interfering others, the SI is trying hard to avoid this while maintaining the system cost especially in the area with air space congestion.

To strike the balance between the cost and performance, the modern pre-configurable antenna shall be equipped with beamforming function and provide the capability to be pre-configured before installation and re-configure during the operation based on the deployment scenario illustrated in the following paragraph. This

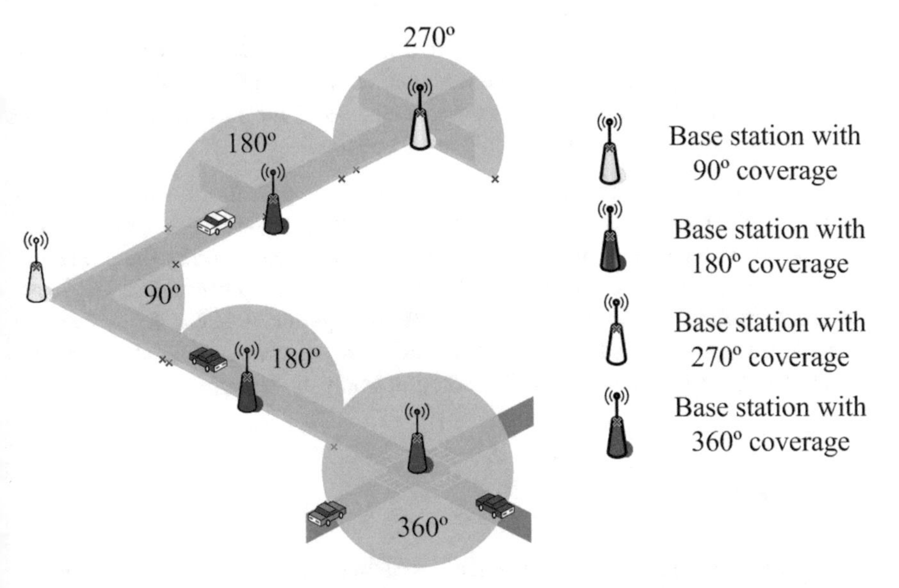

Fig. 2.5 Wireless zone coverage in the vehicular environment for various application scenarios

technique can tremendously improve interference performance, cost reduction, and increase the field deployment friendliness.

In the dynamic transportation environment, specific radiating beam and the communication distance of the antenna [12] may apply to different application scenarios as described in Fig. 2.5, for instance, a big traffic junction (red cell) will require a base station with the entire 360° coverage with maximum antenna gain, a small junction (white cell) may need a base station with 270° wireless coverage with lower antenna gain, a roadside infrastructure installed at the roadside (blue cell) just requires a 180° wireless coverage and a right-angled road (yellow cell) will just require a small 90° beam coverage. To increase the field deployment friendliness further, the gain of the antenna system can be pre-configured to low, middle and high gain before the installation, hence, it can be customised to what is needed to best fit the application scenario. On top of that, the beamforming feature can be realised in the proposed antenna array by integrating the array with the beamforming frontend and processor. The proposed pre-configurable smart antenna structure aimed to achieve the lowest CAPEX and OPEX that fit well in the volatile transportation sector. The proposed pre-configurable antenna system comes with beamforming capability outperformed the conventional smart antenna and reconfiguration antenna system with the following benefits,

(i) The number for equipment needed will be reduced (with beamforming technique)
(ii) Lower cost and minimise waste materials (with pre-configurable structure)
(iii) Maintenance friendly and good maintainability (with fewer components)

(iv) Fewer arrays, less processing effort that resulted in faster processing time. (the beamformer just need to concentrate on the selected arrays) and
(v) Enhanced interference performance (radiate and received on a narrow and concentrated beam on the specific array installed).

It's undeniable that the big chunk of the cost that spent in the wireless infrastructure deployment goes to the equipment and installation cost that was mainly contributed by the number of equipment and the added cost to certify the product in the harsh transportation environment. The natural approach towards cost and interference friendliness is to reduce the number of equipment in the field, this can be achieved via the smart antenna technique with pre-configurable structure and re-configurable beam to focus the RF radiation solely on the targeted directly.

References

1. É. Masson, M. Berbineau, Railway operators needs in terms of wireless communications, in *Broadband Wireless Communications for Railway Applications*. Studies in Systems, Decision and Control, Vol. 82 (Springer, Cham, 2017)
2. W. Chen, *Vehicular Communications and Networks, Architectures, Protocols, Operation and Deployment* (Woodhead, 2015)
3. F. Hu, *Vehicle-to-Vehicle and Vehicle-to-Infrastructure Communications: A Technical Approach* (Taylor & Grancis, 2018)
4. K. Zheng, Q. Zheng, P. Chatzimisios, W. Xiang, Y. Zhou, Heterogeneous vehicular networking: a survey on architecture, challenges, and solutions. IEEE Commun. Surv. Tutor. **17**(4), 2377–2396, Fourthquarter (2015). https://doi.org/10.1109/COMST.2015.2440103
5. V. Kone, H. Zheng, A. Rowstron, G. O'Shea, B.Y. Zhao, Measurement-based design of roadside content delivery systems. IEEE Trans. Mob. Comput. **12**(6), 1160–1173 (2013). https://doi.org/10.1109/TMC.2012.90
6. E. Belyaev, A. Vinel, A. Surak, M. Gabbouj, M. Jonsson, K. Egiazarian, Robust vehicle-to-infrastructure video transmission for road surveillance applications. IEEE Trans. Veh. Technol. **64**(7), 2991–3003 (2015). https://doi.org/10.1109/TVT.2014.2354376
7. J. Heo, B. Kang, J.M. Yang, J. Paek, S. Bahk, Performance-cost tradeoff of using mobile roadside Units for V2X communication. IEEE Trans. Veh. Technol. **68**(9), 9049–9059 (2019). https://doi.org/10.1109/TVT.2019.2925849
8. M.C. Tan, M. Li, Q.H. Abbasi, M. Imran, A smart and low-cost enhanced antenna system for industrial wireless broadband communication, in *12th European Conference on Antennas and Propagation (EuCAP 2018)*, London, 2018, pp. 1–4. https://doi.org/10.1049/cp.2018.1218
9. A. Ghasemi, A. Abedi, F. Ghasemi, Introduction to radiowaves propagation, in *Propagation Engineering in Radio Links Design* (Springer, New York, USA, 2013), Chap. 1, Sect. 1.10.1, pp. 26–27
10. A. Ogunsola, A. Mariscotti, Railway operators needs in terms of wireless communications, in *Electromagnetic Compatibility in Railways: Analysis and Management*, Vol. 168 (Springer Science & Business Media, 2013)
11. smartpres1.pdf, Development of Smart Antenna Technology (2006), https://www.ofcom.org.uk/__data/assets/pdf_file/0014/36014/smartpres1.pdf. Access 29 May 2020
12. M.C. Tan, M. Li, Q.H. Abbasi, M.A. Imran, A wideband beamforming antenna array for 802.11ac and 4.9 GHz in modern transportation market. IEEE Trans. Veh. Technol. **69**(3), 2659–2670 (2020). https://doi.org/10.1109/TVT.2019.2963111

Chapter 3
State-of-the-Art Antenna System

Many research works have been carried out in the beamforming and smart antenna related area, In this book, we are focusing our resources in 3 areas, (i) the smart antenna system, (ii) the phased antenna array and, (iii) the single antenna element that formed the phased antenna array.

3.1 Smart Antenna System

The smart antenna system allows the transmitter and receiver to radiate and receive it RF signal from the calculated direction with the best reception quality and lowest operating beamwidth to avoid mutual interference with the neighboring devices. The beam switching method for the conventional smart antenna that covers 360° azimuth plane can be explained in Fig. 3.1, the antenna was designed to operate at a fixed number of beams and each beam occupied the fixed beamwidth, for instance, a 6 switched beams antenna will have a 60° beamwidth on each beam. During the smart antenna operation, one of the beams will be activated at one time according to the targeted direction of the transmitter and receiver.

In the smart antenna literature, most of the works were concentrated around the reconfigurable structure and beamforming antenna, in this book, we are looking into the issue from the angle of usage scenarios and practical application that combines multiple phased arrays with wide operating band and beamforming capability to address the different application scenarios in the dynamic transportation sector as described in Fig. 2.5. In the past, a reconfigurable polarization beamforming antenna [1] that operates in 5 radiating beams covering the whole 360° with 3 polarizations in the 2.4 GHz band was proposed, the reconfigurable was realized by changing the bias voltage of the Positive Intrinsic Negative (PIN) diodes on the parasitic elements and the polarized reconfigurable square patch, the realized gain for the 5 beams are between 2.5 and 3.5 dBi.

M. C. Tan et al., *Antenna Design Challenges and Future Directions for Modern Transportation Market*, SpringerBriefs in Applied Sciences and Technology,
https://doi.org/10.1007/978-3-030-61581-9_3

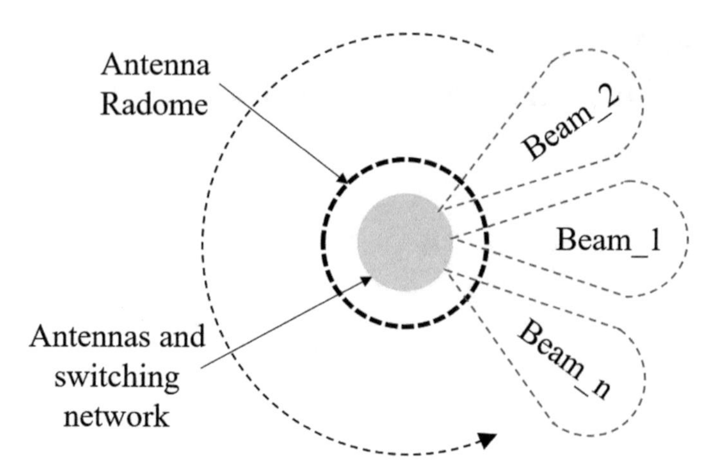

Fig. 3.1 Beam switching method for the smart antenna that covers 360° azimuth plane

Similarly, in [2], the authors have proposed a reconfigurable beam switching antenna with dual-polarization, the design was constructed by radial waveguide as center probes surrounded by radiating elements, the polarization and the radiating beam are controlled by 36 pieces of PIN diodes, the proposed antenna was targeted for the wireless local area network (WLAN) at 5 GHz band between 5.18 and 5.825 GHz. The prototype was tested with the realized gain of around 12 dBi and the HPBW of around 30° that allows the antenna to switch among the 12 fixed beams within the 360° azimuth plane.

The smart antenna that can be reconfigured by beam switching has been proposed In [3] making use of a patch as radiating element and six parasitic elements that can be controlled by the PIN diodes, the design was able to achieve the beam switching up to six directions over 360° with the azimuth beamwidth of 42° and the peak gain of around 10 dBi. Another reconfigurable switched beam antenna system [4] was designed using the slotted array that is capable of performing beam scanning in the elevation plane, the design was targeted for LTE femtocells, the HPBW was around 25° with the down-tilt angles of 13° and 32° that capable to switch between 3 elevation beams (62°, 82° and 108°) via the series-fed structure. The achievable antenna gain was around 9.9 dBi.

In [5], the Electronically Steerable Parasitic Array Radiators (ESPARs) has been proposed, the parasitic elements consist of 12 folded monopole antennas which were surrounded by a short monopole antenna at the middle of the antenna structure, each folded monopole was controlled by a PIN diode and the driven element was implemented using the short monopole antenna, the beam switched over the 360° in the azimuth plane was configured via the PIN diodes. The operating frequency of the antenna was from 1.8 to 2.2 GHz and 2.85 to 3.15 GHz with the realized gain of 6.5 for 2 GHz and 5.5 dBi for 2.9 GHz. The combination of PIN diode switching can generate up to 12 beams with 30° beams steering resolution.

Another 360° beam-steering antenna was proposed with gain enhancement and size reduction by adopting the Yagi-Uda arrangement using the small director array (SDA) [6], the reconfigurable was achieved using switched parasitic elements via the PIN diode to steer the beam from 0° to 360° in the azimuth plane. The antenna operates from 2.4 to 2.5 GHz with 10 dBi measured gain. The antenna was designed to operate up to 6 beams with 60° beam-steering resolution.

A 360° beam-steering antenna has been implemented by combining a monopole and patch as simultaneous radiation elements as reported in [7], the antenna consists of two ports, one connected to the diploe and the other to the patch. By injecting the RF signal to the ports with different phase, 4 beams can be selected over the entire 360° azimuth plane, each selected beam comes with two symmetrical beams. The proposed antenna operates from 2.235 to 2.725 GHz and the measure gain is between 2.1 and 3.1 dBi.

The reconfigurable techniques in [1–6] are the popular techniques adopted for beam steering antenna, where the PIN diodes were used to switch the radiating beam of the antenna, however, such technique has its limitations due to the fixed beams and wide beamwidth that limit the antenna system for the operations that required narrower and fine resolution on the steering angle, as the technology migrates into digital beamforming that requires more precise beamforming angle, fine beam-steering resolution and small beamwidth to fulfil the high interference rejection requirement.

3.2 Antenna Array

The phased array plays an important part in our proposed smart antenna system, the choice of the antenna array design will determine the smart antenna performance such as operating bandwidth, gain, sidelobe level etc. Antenna arrays design seems to be a popular topic in the literature, however, the applications were mainly focused on the area such as 5G [8, 9–13], X-band [14], LTE [15], 2.4 GHz ISM band [16] and 2.35–2.8 GHz/5–5.5 GHz [17]. In the Wi-Fi arena such as 802.111/an/ac, the literature was mainly focused on array miniaturization, enhance the radiation performance and structure and leave with limited work on enabling the array to operate in a wider band that covers the 4.9 GHz licensed band and 5.1–5.9 GHz ISM band. For instance, in [18], a 2×4 dual-polarization antenna array that operates from 5.15 to 5.85 GHz with the realized gain of 12 dBi was proposed. In another work [19], the Complementary Split Ring Resonators (CSRR) method was used to reduce the size of the 5 GHz patch antenna, however, the gain suffered at only −0.16 dBi and limited operating bandwidth around 80 MHz, similarly, in [20], the 8 elements MIMO antenna with 6.5 dBi gain and narrow operating frequency bandwidth from 4.985 to 5.15 GHz has been proposed. Furthermore, in [21], a 3 monopole patches antenna array with inverted L and inverted Z came with slightly wider bandwidth covering from 5 to 6 GHz and achievable gain from 2 to 3.7 dBi was proposed. In this work, a wideband and high-gain phased array that operates from 4.9 to 5.9 GHz band will be introduced, the proposed wideband array will be further integrated into the pre-configurable smart antenna system.

3.3 Antenna Elements

The type of antenna elements in the phased array will determine the performance and radiation characteristic of the phased array. One of the popular antenna element that widely used in the phased array design is Microstrip Patch Antenna (MPA) [22], the reason was due to the flexible MPA structure, simple to manufacture and lower in cost. The flexible structure allows easy customization and performance enhancement such as gain and bandwidth improvement. The MPA has been enhanced through various methods such as altering the structure of the MPA, changing the material of the dielectric substrate, different feeding methods etc., as a result, the gain and operation bandwidth limitation have been improved over the years. A 14% bandwidth ratio wideband multilayer patch antenna structure was proposed in [23], however, the design may not be suitable for large array integration due to the nature of the stacked capacitive coupling through air structure that may not be manufacture friendly. An open slot was introduced on the radiating patch in [24] to enhance the bandwidth ratio to 22% but it suffered a lower gain of around 3.25 dBi. The Electromagnetic BandGap (EBG) [25] and Defective Ground Structure (DGS) [26] techniques have delivered the bandwidth ratio of 18.68% and 63.65% respectively with the bigger element size. An ultra-wideband dual substrate MPA with a small capacitive feed was proposed in [27], the bandwidth ratio was around 50% and exceeding 7 dBi gain, the capacitive coupling was realized by the spacing between the radiating patch and the feeding pad on the same PCB layer that increases the manufacturability of the antenna, the dual substrate capacitive feed MPA elements can be a potential candidate for further integration and optimized into a beamforming array.

3.4 The Modern Pre-Configurable Smart Antenna System

The conventional smart antenna was aimed to combat the sticky interference issue due to air space congestion. In our case, we are aiming to further enhance the smart antenna by taking into consideration various scenarios in the practical deployment [28]. The idea of this smart beamforming antenna systems is aimed to improve the wireless communication in the transportation environment in 3 areas, (i) **performance**, interference and range performance can be improved by concentrating the antenna transmit power and receiving to the direction of the target and nulls for other direction, (ii) **cost**, the range coverage improvement has reduced the number of base stations needed, in addition, the pre-configuration method allows only the necessary sub-array to be installed based on the coverage and range requirement, hence it saves the material cost, (iii) **maintainability**, with the less number of equipment installed in the field that implies fewer resources and cost is needed to maintain the system throughout the operating years. Our design is aiming for an antenna structure with a wide operating frequency band that can be pre-configured to different coverage sector and gain with narrow beam that suports beamforming feature within the sector, the proposed pre-configurable smart antenna structure is expected to improve the interference rejection and deployment cost in the dynamic transportation environment. The smart antenna system is expected to deliver the following features,

a. A Wide operating band that covers the 4.9 GHz licensed band and 5.1–5.9 GHz ISM band that can be materialized by the dual-substrates technique by integrating the MPA on two thick substrates consist of air and the F4BTM-2 substrate.
b. Different array with choice of gain can be pre-selected from the prefabricated 1 × 4, 2 × 4 and 4 × 4 arrays, for example, 11.16, 14.59 and 17.25 dBi.
c. Pre-configurable sector (each sector covers 90° scanning angle) can be pre-configured to enable coverage over the 90°/180°/270°/360° area, up to 4 arrays to support 360° coverage.
d. 24° beamwidth with beam steering function.

The illustrations of the pre-configurable smart antenna system are presented in Figs. 3.2 and 3.3. The phased array was designed with multiple capacitive feed MPA elements lay over the structure of the dual substrates to achieve the higher gain and wider operating bandwidth. The volatile antenna structure enables the pre-configurable ability of the smart antenna systems to tackle the needs to have the configurable gain and beam coverage angle in order to obtain the optimum setup that suite well in the dynamic vehicular environment, with different physical coverage beam and communication range can be pre-configured for different application scenarios as described earlier in Fig. 2.5.

Firstly, the antenna element was designed with high gain and wide operating bandwidth and constructed with the feed network into the modular sub-arrays, 3 types of modular linear sub-arrays that support 90° beam steering were pre-designed and

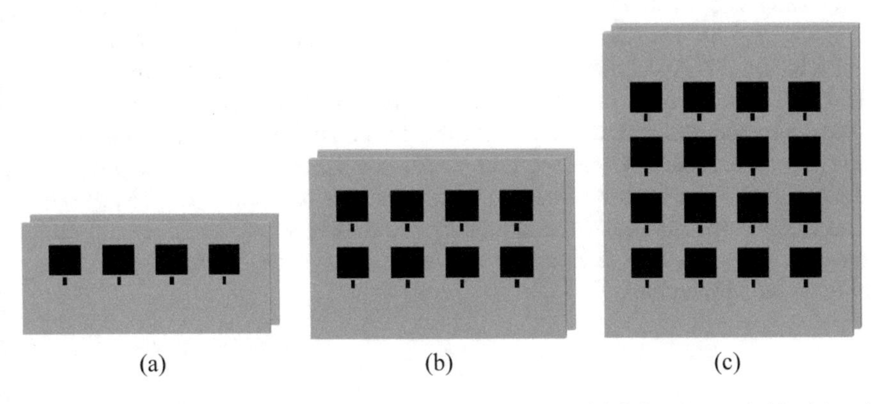

Fig. 3.2 Arrays with pre-configurable gain. **a** 1 × 4 array (low gain). **b** 2 × 4 array (mid gain). **c** 4 × 4 array (high gain)

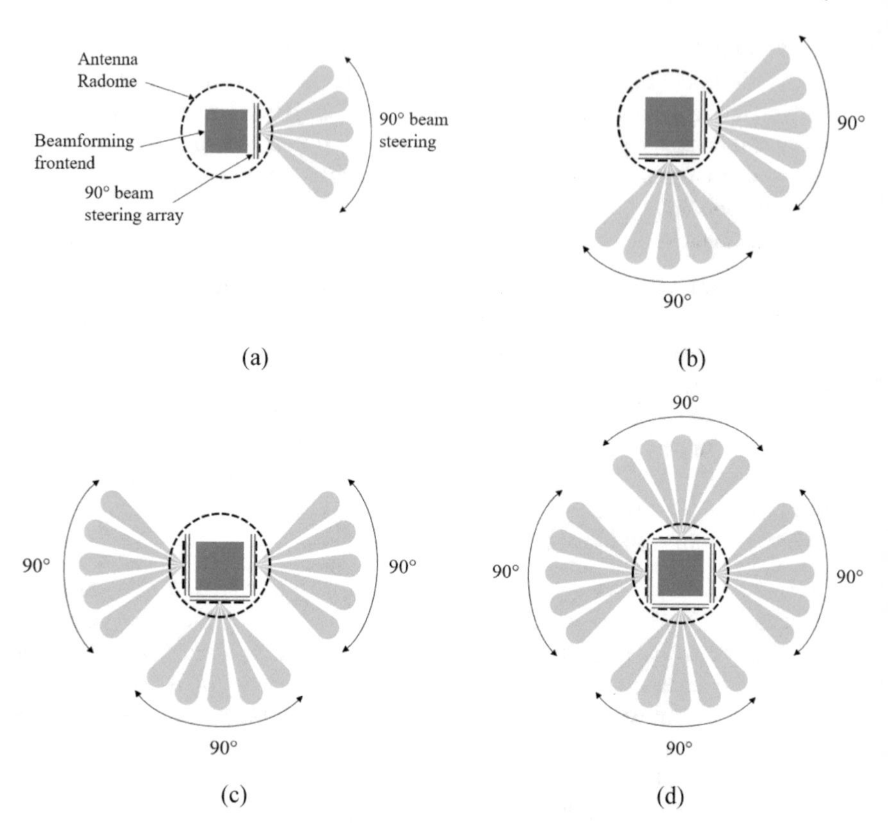

Fig. 3.3 Pre-configurable antenna system. **a** 90° sector with one sub-array. **b** 180° sector with 2 sub-arrays. **c** 270° sector with 3 sub-arrays. **d** 360° sector with 4 sub-arrays

fabricated with low, medium and high gain as shown in Fig. 3.2, and then following the 2 simple configuration steps to form the 360° pre-configurable antenna system. The first step allowing the user to pick the desired gain from the modular sub-arrays listed in Fig. 3.2. And the second step is shown in Fig. 3.3, the beam coverage of 90°, 180°, 270°, and 360° can be constructed by simply integrate the multiple sub-arrays onto the antenna systems; the similar method can be used to configure the middle gain and high gain antenna system, the cone shapes highlighted in orange colour represent the beams of the sub-array which can be electronically steered within the 90° sector. The space at the middle of the radome is reserved to house the RF beamforming frontend that consists of electronic components mounted on the PCB to perform and beam steering function.

References

1. C. Gu et al., Compact smart antenna with electronic beam-switching and reconfigurable polarizations. IEEE Trans. Antennas Propag. **63**(12), 5325–5333 (2015). https://doi.org/10.1109/TAP.2015.2490239
2. H. Boutayeb, P.R. Watson, W. Lu, T. Wu, Beam switching dual polarized antenna array with reconfigurable radial waveguide power dividers. IEEE Trans. Antennas Propag. **65**(4), 1807–1814 (2016). https://doi.org/10.1109/TAP.2016.2629469
3. Y. Yang, X. Zhu, A wideband reconfigurable antenna with 360° beam-steering for 802.11ac WLAN applications. IEEE Trans. Antennas Propag. **66**(2), 600–608 (2017). https://doi.org/10.1109/TAP.2017.2784438
4. Y.Y. Lin, C.L. Liao, T.H. Hsieh, W.J. Liao, A novel beam switching array antenna using series-fed slots with PIN diodes. IEEE Antennas Wirel. Propag. Lett. **16**, 1393–1396 (2016). https://doi.org/10.1109/LAWP.2016.2639046
5. H. Liu, S. Gao, T.H. Loh, Compact dual-band antenna with electronic beam-steering and beamforming capability. IEEE Antennas Wirel. Propag. Lett. **10**, 1349–1352 (2011). https://doi.org/10.1109/LAWP.2011.2177059.7115909
6. H. Liu, S. Gao, T. Loh, Small director array for low-profile smart antennas achieving higher gain. IEEE Trans. Antennas Propag. **61**(1), 162–168 (2013). https://doi.org/10.1109/TAP.2012.2219841
7. A. Narbudowicz, M.J. Ammann, D. Heberling, Switchless reconfigurable antenna with 360° steering. IEEE Antennas Wirel. Propag. Lett. **15**, 1689–1692 (2016). https://doi.org/10.1109/LAWP.2016.2524199
8. S. Krishna, G. Mishra, S. Shama, A series fed planar microstrip patch array antenna with 1D beam steering for 5G spectrum massive MIMO applications, in *IEEE Radio and Wireless Symposium*, Anaheim, CA, USA, 2018, pp. 209–212
9. R.R. Selvaraju, M.R. Kamarudin, M.H. Jamaluddin, M.H. Dahri, C.Y. Low, Compact 4-element beam steerable printed adaptive array antenna for 5G application, in *Asia-Pacific Conference on Applied Electromagnetics (APACE)*, Langkawi, Malaysia, 2016, pp. 30–33
10. Y. Wang, H. Wang, G Yang, Design of dipole beam-steering antenna array for 5G handset applications, in *Progress in Electromagnetic Research Symposium (PIERS)*, Shanghai, China, 2016, pp. 2450–2453
11. M. Mantash, T.A. Denidni, Millimeter-wave beam-steering antenna array for 5G applications, in *IEEE 28th Annual International Symposium on Personal, Indoor, and Mobile Radio Communications (PIMRC)*, Montreal, QC, Canada, 2017, pp. 1–3
12. J.H. Kim, J.H. Han, J.S. Park, J.G. Kim, Design of phased array antenna for 5G mm-wave beamforming system, in *IEEE 5th Asia-Pacific Conference on Antennas and Propagation (APCAP)*, Kaohsiung, Taiwan, 2016, pp. 201–202
13. Y. Wang, L. Zhu, H. Wang, Y. Luo, G. Yang, A compact, scanning tightly coupled dipole array with parasitic strips for next-generation wireless applications. IEEE Antennas Wireless Propag. Lett. **17**(4), 534–537 (2018). https://doi.org/10.1109/LAWP.2018.2798660
14. S.A. Aghdam, J. Bagby, R.J. Pla, Design and development of linear array of rectangular aperture coupled microstrip antennas with application in beamforming, in *17th International Symposium on Antenna Technology and Applied Electromagnetics (ANTEM)*, Montreal, QC, Canada, 2016, pp. 1–3
15. J. Nasir, M.H. Jamaluddin, M.R. Kamarudin, Irfanullah, Y.C. Lo, R. Selvaraju, A four-element linear dielectric resonator antenna array for beamforming applications with compensation of mutual coupling, IEEE Access **4**, 6427–6437 (2016). https://doi.org/10.1109/ACCESS.2016.2614334
16. D. Pavithral, P. Ramya, K.R. Dharani, M.R. Devi, Design of microstrip patch array antenna using beamforming technique for ISM band, in *Fifth International Conference on Advanced Computing (ICoAC)*, Chennai, India, 2013, pp. 504–507

17. M.S.R. Bashri, T. Arslan, W. Zhou, A dual-band linear phased array antenna for WiFi and LTE mobile applications, in *Loughborough Antennas and Propagation Conference (LAPC)*, Loughborough, UK, 2015, pp. 1–5
18. C.Y. Cheng, H.Y. Huxie, F.H.L. Su, A compact high gain patch antenna array for IEEE 802.11ac MIMO application, in *IEEE 5 th Asia-Pacific Conference on Antennas and Propagation*, Kaohsiung, Taiwan, 2016, pp. 327–328
19. M.U. Khan, M.S. Sharawi, A compact 8-element MIMO antenna system for 802.11ac WLAN applications, in *International Workshop on Antenna Technology (iWAT2013)*, Karlsruhe, Germany, 2013, pp. 91–94
20. S.A. Nasir, M. Mustaqim, B.A. Khawaja, Antenna array for 5th generation 802.11ac Wi-Fi applications, in *11th Annual High Capacity Optical Networks and merging/Enabling Technologies (Photonics for Energy)*, Charlotte, NC, USA, 2014, pp. 20–24
21. W.S. Chen, C.Y. Hsu, F.S. Chang, Broadband three-element MIMO antennas for IEEE802.11ac, in *IEEE 5th Asia-Pacific Conference on Antennas and Propagation (APCAP)*, Kaohsiung, Taiwan, 2016, pp. 267–268
22. C.A. Balanis, Microstrip and mobile communications antennas, in *Antenna Theory Analysis and Design*, 4th edn. (Wiley, Hoboken, New Jersey, USA, 2016), Chap. 14, Sect. 14.2, pp. 788–823
23. G. Giunta, C. Novi, S. Maddio, G. Pelosi, M. Righini, S. Selleri, Efficient tolerance analysis on a low cost, compact size, wideband multilayer patch antenna, in *IEEE International Symposium on Antennas and Propagation and USNC/URSI National Radio Science Meeting*, San Diego, CA, USA, 2017, pp. 2113–2114
24. K. Mandal, L. Murmu, P.P. Sarkar, Investigation on compactness, bandwidth and gain of circular microstrip patch antenna, in *IEEE Devices for Integrated Circuit (DevIC)*, Kalyani, India, 2017, pp. 742–746
25. R. Gupta, M. Kumar, Bandwidth enhancement of microstrip patch antennas by implementing circular unit cells in circular pattern, in *5th International Conference on Computational Intelligence and Communication Networks*, Mathura, Uttar Pradesh, India, 2013, pp. 10–13
26. S.N. Ather, R.K. Verma, P.K. Singhal, Bandwidth enhancement for truncated rectangular microstrip antenna using stacked patches and defected ground structure, in *5th International Conference on Computational Intelligence and Communication Networks*, Mathura, India, 2013, pp. 55–56
27. V.G. Kasabegoudar, D.S. Upadhyay, K.J. Vinoy, Design studies of ultra-wideband microstrip antennas with a small capacitive feed. Int. J. Antennas Propag. (2007). https://doi.org/10.1155/2007/67503
28. M.C. Tan, M. Li, Q.H. Abbasi, M.A. Imran, A wideband beamforming antenna array for 802.11ac and 4.9 GHz in modern transportation market. IEEE Trans. Veh. Technol. **69**(3), 2659–2670 (2020). https://doi.org/10.1109/TVT.2019.2963111

Chapter 4
Designing the Smart Antenna System

This section will run through the necessary steps towards designing the smart antenna system, starting from the type of antenna element selection, single element design with enhanced operating bandwidth, gain and parameters optimisation, followed by the ground plane simulation to identify the ground plane effect, the best elements separation can be determined so that the optimum gain, beamwidth and sidelobe performance can be achieved. Following next is to combine the vertical elements into a single feeding port via the microstrip feeding network, the elements are arranged symmetrically to form a linear array with 4 RF ports that allow beamforming function to take place. 3 types of arrays 1×4, 2×4 and 4×4 were designed.

4.1 The Microstrip Patch Antenna Elements

Among many types of antenna element structure, the MPA structure is the popular type of element, it's simple to design and easily integrate into antenna array that supports higher gain, the small form factor of the MPA also make it popular for application in mobile devices, vehicular application and long-range communication antennas. The MPA structure consists of four parts (patch, port feed, substrate, and ground plane). A thin copper patch is laid over a non-conductive dielectric substrate with the finite ground plane at the opposite side of the substrate. The patch is made of copper foil coated with anti-corrosion material such as gold, the substrate materials use in PCB usually consists of fibreglass with a dielectric constant between 2 and 4.5. Copper foil or solid metal plate can be used as a ground plane depends on the type of application.

Over the year, many shapes of the MPA were designed, and the most popular shapes are the rectangular and circular shape [1]. The design consideration for the circular patch is relatively simple compared to rectangular patch, for example, the circular patch has only one degree of freedom, which is to control the radius of the

M. C. Tan et al., *Antenna Design Challenges and Future Directions for Modern Transportation Market*, SpringerBriefs in Applied Sciences and Technology,
https://doi.org/10.1007/978-3-030-61581-9_4

MPA patch, however, the rectangular patch has two degrees of freedom to control the length and width of the MPA patch. The characteristic of the conventional rectangular and circular shape MPA without any gain and bandwidth enhancement was reproduced and simulated using CST tool, the parameter of the rectangular and circular MPA are presented in Figs. 4.1 and 4.2 and the result for 5.5 GHz operating frequency are tabulated in Table 4.1, the conventional MPA antenna normally comes with a narrower operating bandwidth around 4% and lower gain of less than 5 dBi. In practical application, many works and techniques have been carried out to overcome

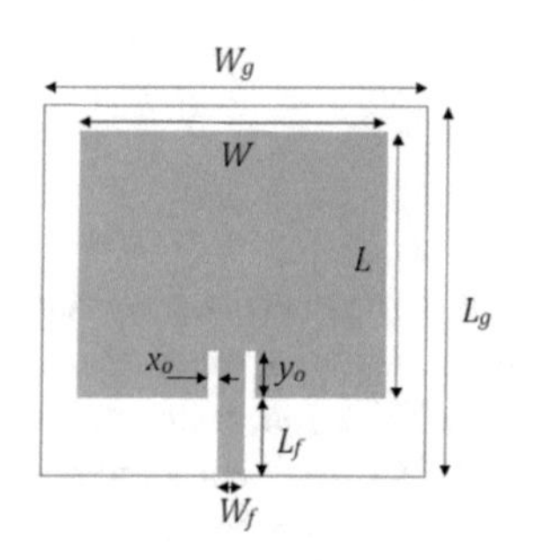

Where:
Wg: width of the ground plane/Substrate 26.3535 mm
Lg: length of the ground plane/Substrate 22.1795 mm
W: width of the MPA Patch 16.7535 mm
L: length of the MPA Patch 12.5795 mm
Lf: length of the inset feed 8.8 mm
Wf: width of the inset feed 2.927 mm
xo: the gap between the antenna and inset feed 1 mm
yo: length of the inset feed 4 mm
tp: thickness of the patch 0.2 mm
hs: height of the substrate 1.6 mm
εr: relative permittivity of the dielectric substrate 4.3

Fig. 4.1 Conventional rectangular MPA antenna element

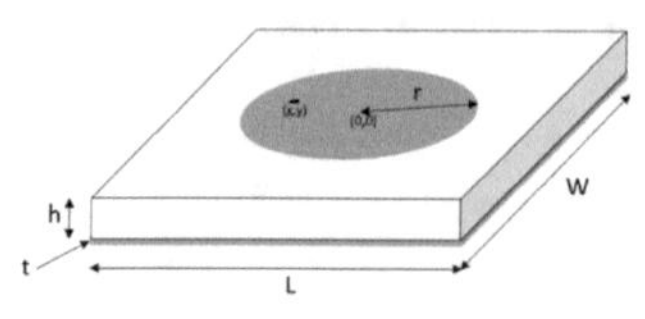

Where:
r: patch radius 7.25 mm
t: patch thickness 0.1 mm
h: substrate thickness 1.6 mm
x: feeding point, x 2.625 mm
y: feeding point, y 0 mm
L: length of the ground plane 29 mm
W: width of the ground plane 29 mm
ε_r: relative permittivity of the substrate material 4.3

Fig. 4.2 Conventional circular MPA antenna element

Table 4.1 Conventional MPA antenna performance (without bandwidth and gain enhancement)

Characteristic	Conventional rectangular patch	Conventional circular patch
Frequency	5.573 GHz	5.336 GHz
Operating band	5.47–5.69 GHz	5.21–5.46 GHz
Bandwidth	0.22 GHz	0.25 GHz
Bandwidth ratio	3.94%	4.7%
Gain	1.76 dBi	4.78 dBi
Gain characteristic (4.9–5.6 GHz)	−5.42 to 1.69 dBi	0.24–4.54 dBi
In-band gain ripple	7.11 dB	4.3 dB

the limitation of the MPA to make it useable, performance parameters such as gain, operating bandwidth, in-band gain ripple etc. have been improved other the years.

4.2 Designing Antenna Element and Optimisations

The antenna array was built by multiple single antenna elements placed in the plane vertically and horizontally. The single elements constructed with the dual substrate and feed via capacitive coupling techniques in [2] was chosen in our design. The large patch as shown in Fig. 4.3 act as the radiator and the small patch function as a capacitive feeding network to allow the energy feed through from the coaxial feed to the radiating patch. The potential mismatch may occur between the co-axial feeding cable and the patch can be reduced by the capacitive feed method, the bandwidth and gain of the MPA antenna can be enhanced by the 2 layers of substrate. The radiating patch of the antenna was designed on an F4BTM-2 substrate with relative permittivity, $\varepsilon_r = 3$, to resonate at 5.5 GHz. The length, L and width, W of the patch were calculated using [3],

$$W = \frac{1}{2f_r\sqrt{\mu_o\varepsilon_o}}\sqrt{\frac{2}{\varepsilon_r+1}} = \frac{c}{2f_r}\sqrt{\frac{2}{\varepsilon_r+1}} \tag{4.1}$$

where,

W: the width of the patch
f_r: the center frequency
μ_o: magnetic permeability of free space
ε_o: electric permeability of free space
ε_r: dielectric constant of the substrate
c: speed of light.

$$L = \frac{c}{2f_r\sqrt{\varepsilon_{reff}}} - 2\Delta L \tag{4.2}$$

where,

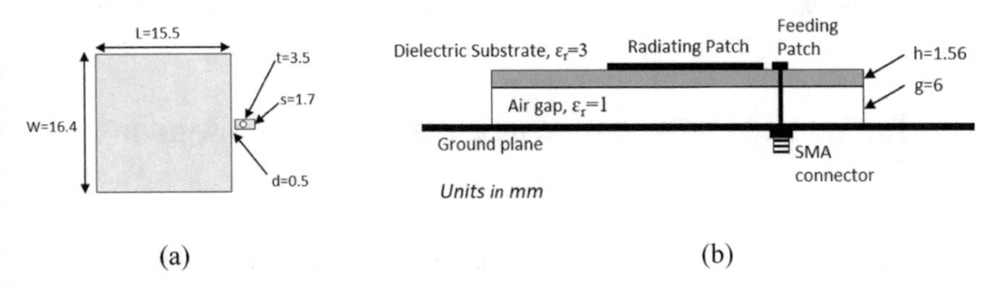

Fig. 4.3 Capacitive feed dual-substrate MPA. **a** Top view. **b** Side view

L: the length of the patch
ε_{reff}: the effective dielectric constant
ΔL: the extended length of the patch that can be calculated using Eqs. (4.3) and (4.4).

$$\varepsilon_{reff} = \frac{\varepsilon_r+1}{2} + \frac{\varepsilon_r-1}{2}\left[1 + 12\frac{h}{W}\right]^{-\frac{1}{2}} \tag{4.3}$$

where h is the height of the substrate.

$$\Delta L = 0.412h\frac{(\varepsilon_{reff}+0.3)\left(\frac{W}{h}+0.264\right)}{(\varepsilon_{reff}-0.258)\left(\frac{W}{h}+0.8\right)} \tag{4.4}$$

The optimum operating frequency bandwidth can be optimised by adjusting the capacitive feed length (t), width (s), distance to patch (d), and the space of airgap (g). In this work, we are making use of the design in [2] and further optimize through parameter optimization to improve the frequency bandwidth to cover our interest band from 4.9 to 5.9 GHz as well as enhancing the antenna gain and cap the returned loss to below 10 dB, from the simulation result, about 15% improvement was achieved on both operating bandwidth and gain compared to the original design parameter in [2].

The antenna parameters were optimized and simulated using the Computer Simulation Technology (CST) Studio Suite (Darmstadt, Germany) [4], the optimized parameters are shown in the geometrical model in Fig. 4.3. The simulation results of the single capacitive feed MPA element are presented in Fig. 4.4. The returned loss of 10 dB was achieved between 4.73 and 7.09 GHz which translate to around 43% bandwidth ration at 5.5 GHz center frequency. The maximum gain of 8.6 dBi at 5.5 GHz was achieved with the gain flatness of 0.8 dB over the 4.9–5.9 GHz band, which is quite reasonable for the entire 1 GHz bandwidth.

In the phased array that was constructed by arranging multiple MPA elements to form a bigger array that directly contributes to the increase in ground plane size, the performance impact is evaluated by observing the return loss of the element with respect to the ground plane size, the simulation results are captured in Fig. 4.5, the results have proofed that the variations of ground plane dimension has minimum impact to the antenna performance, thus we concluded that the antenna element can be used to design into a phased array that usually occupied a big ground plane.

4.3 Design and Optimization of the Beamforming Antenna Array

Phased arrays are used to synthesize a specific radiating pattern that cannot be achieved with a single element antenna [1], to achieve that, the spacing between

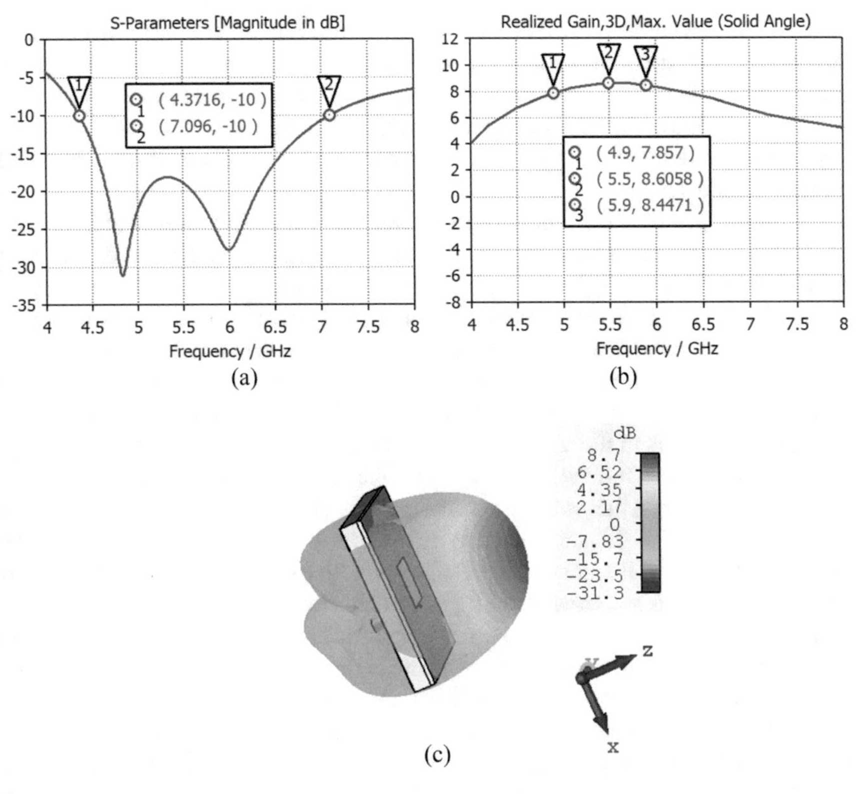

Fig. 4.4 **CST** Simulation results of the capacitive feed MPA element. **a** Simulated S11 result. **b** Gain characteristic over the designed band. **c** Far-field radiation pattern

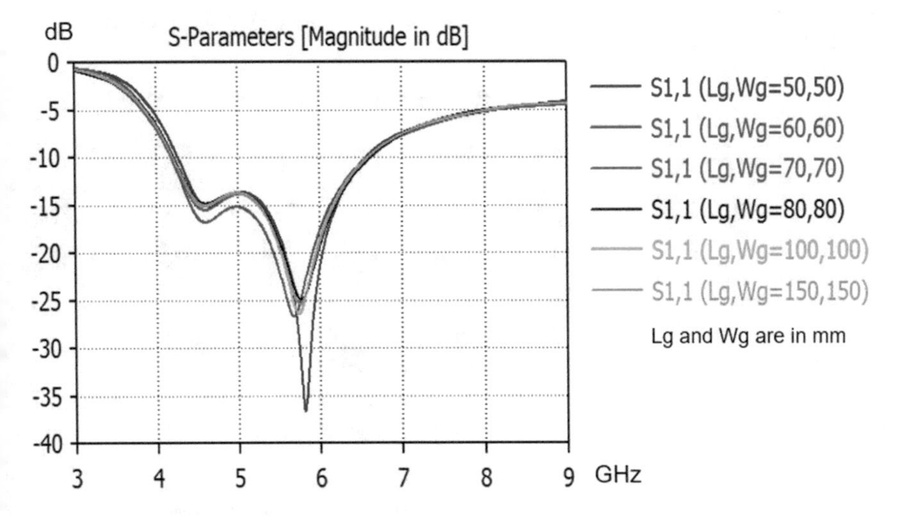

Fig. 4.5 **Simulation on the** effect of the size of the ground plane. *Note Lg is the length of the ground plane and Wg is the width of the ground plane*

the element needs to be properly defined and optimized in order to achieve optimum performance in term of gain, sidelobe, and beamwidth. Before the integration of the single element into the bigger array, the proposed 4 × 4 array as shown in Fig. 4.6 was first evaluated using CST simulations tool [4]. The gain, sidelobe and beamwidth performance with respect to the element separation of the antenna array is presented in Figs. 4.7 and 4.8.

Figure 4.7 presents the relationship between the operating beamwidth and the gain of the 4 × 4 phased array with respect to the element separation, as the element separation increases, the gain will gradually increase until it reaches the maximum gain of 21 dBi at 46.025 mm element separation or slightly less than 1λ (wavelength), the gain became saturated when the element separation increases beyond 1λ. Separately, the operating beamwidth became narrower in the exponential rate with the increases in the element separation, for instance, the beamwidth at 20 mm element separation decreases in the rate of 30% for 10 mm increase in the element spacing, the rate of the beamwidth decreases is lower when the element separation increases. The trade-off between the gain and operating beamwidth seems straight forward and easy to understand if the antenna array is used as a point to point link where the gain is maximised while maintaining the beamwidth to as narrow as possible, when comes to beamforming array, we will need to consider another factor which is the sidelobe level (SLL), usually, the best SLL appear at 0° beam, however, it tends to increase when the beam is steered away from its 0° beam, hence, the SLL is another contributor to the performance trade-off that required much attention.

The impact of elements spacing to SLL performance when the radiating been steer from 0° to 30° is presented in Fig. 4.8. For evaluation purposes, the performance of the 30° beam is tested to demonstrate the SLL impact due to the steered beam, and

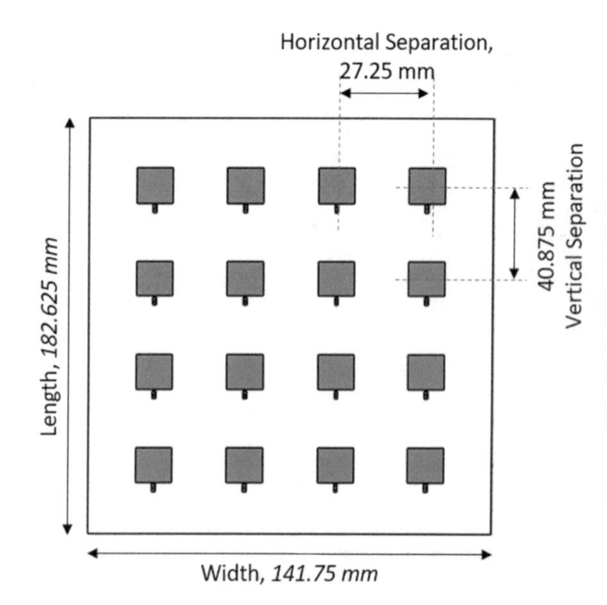

Fig. 4.6 The 4 × 4 array Illustration for "Far Field" simulation

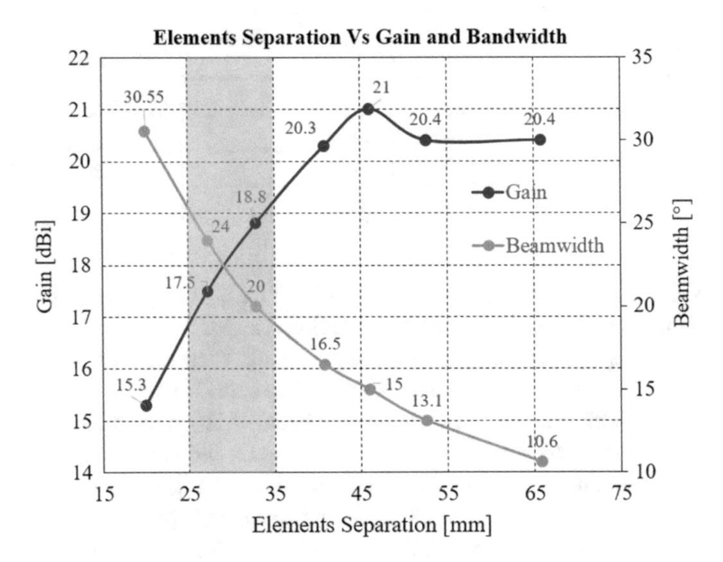

Fig. 4.7 The impact of element spacing to array gain and beamwidth

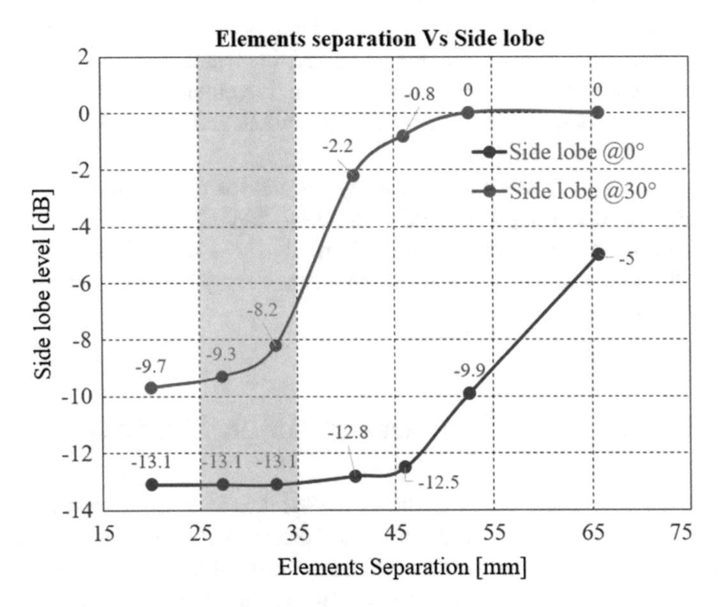

Fig. 4.8 The impact of element spacing to the sidelobe level at 0° and 30° beam

similar effects are applied for other angles. As expected, the lowest SLL level of −13.1 dB is observed at the 0° beam, and the SLL start to increase rapidly when the element spacing increased beyond 45 mm. In the case of 30° beam, the best SLL observed was around −9.7 dBm and the SLL increases rapidly in the higher

Table 4.2 Simulation result of the 4 × 4 arrays using the proposed element spacing

RF parameters	Performance
Gain	17.5 dBi
Operating beamwidth	24°
SLL (at 0° beam)	−13.1 dB
SLL (at 30° beam)	−9.3 dB

rate when the element spacing increases. For both cases, the grating lobe occurred with 2 end-fire maxima when the element spacing reached around 1λ. Therefore, we concluded that the optimum element separation should keep between 25 and 35 mm. This will ensure optimum gain, beamwidth and SLL performance.

Thus far, the performance trade-off such as beamwidth, gain and SLL is demonstrated. As the proposed smart antenna is focusing on the azimuth beam steering, the possible element spacing in Figs. 4.7 and 4.8 are highlighted in green, we choose 27.25 mm (0.5λ) for horizontal separation which produce optimum gain and SLL performance, 40.875 mm (0.75λ) for vertical separation that gives the good beamwidth and gain performance as well as better mutual coupling between the vertical spaced elements. The performance of the array is further simulated using the proposed element spacing, promising results were observed and tabulated in Table 4.2. For other application such as a point-to-point link that usually targeted for higher gain and does not require beamforming function, the element spacing can be set to 45 mm, that will produce 21 dBi gain, 15° beamwidth and −12.5 dB SLL, which is reasonably good for such application.

The 1 × 4 configurations of antenna array designed using the similar MPA configuration was presented in [5] that operates in 4.19–6.58 GHz band with the bandwidth ratio of 44.38%, the 1 × 4 array provides 12.6 dBi gain, −12.9 dB SLL at 0° steering beam and 24.4° beamwidth, in addition, the antenna array is also capable to perform ±40° beam steering.

4.4 Designing the Feeding Network for the Phased Array

Now, we have fixed the spacing of the elements, next is to integrate the vertically spaced elements into a single feed point via the microstrip feeding network. The n × 4 beamforming array is formed by arranging the 4 columns of the vertical arrays horizontally. There are 2 types of popular microstrip feeding networks for MPA array design, they are single line feed (series feed) and multiple lines feed (corporate feed) [1] as shown in Fig. 4.9. The simplest way is to use series feed, however, the drawback is mutual performance effect such as elements coupling and microstrip line reflections if changes made to the feed line or antenna element. For beamforming phased array, corporate feed method is recommended which is less susceptible to design changes, though it required a little bit more space to implement the feeding,

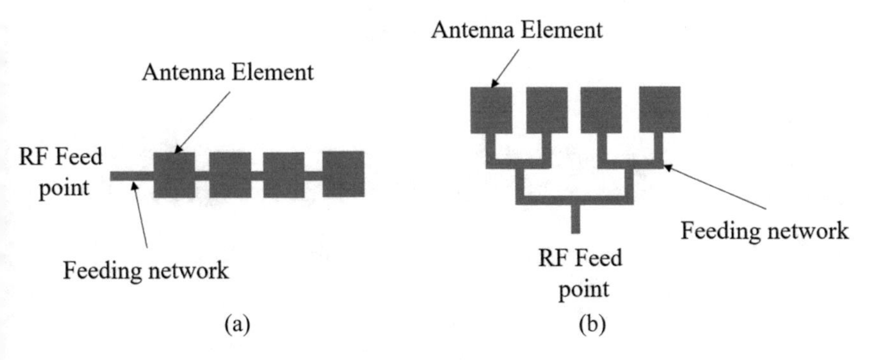

Fig. 4.9 Antenna feeding networks. **a** Series feed. **b** Corporate feed

however, the concern can be overcome by a good PCB layout techniques to lay the feeding line between the available space provided by the element spacing.

In this work, the corporate feed method will be adopted. The n numbers of vertical elements (for instance n = 2 for 2 × 4 array and n = 4 for 4 × 4 array) are joint at the bottom side of the bottom PCB via the microstrip line with 1/4λ impedance transform between the element and the feeding co-axial. The geometry view of the corporate feed array is shown in Fig. 4.10.

The proposed antenna array is designed using the capacitive feed MPA with stack-up structure combining the air and PCB substrate discussed earlier in Sect. 4.2. The radiating elements are located at the top side of the top PCB using F4BTM-2 with $\varepsilon_r = 3$ as substrate, the air gap sandwiched within the radiating PCB (top) and the grounding/feeding PCB (bottom) to act as the second substrate. The bottom PCB consists of a solid ground plane on the top layer that will act as the reflector to the radiating element as well as providing a good mutual coupling rejection between the radiating element and the feeding line which is located at the opposite side of the solid ground plane. The details cross-sectional view of the PCB stack-up construction is captured in Fig. 4.11.

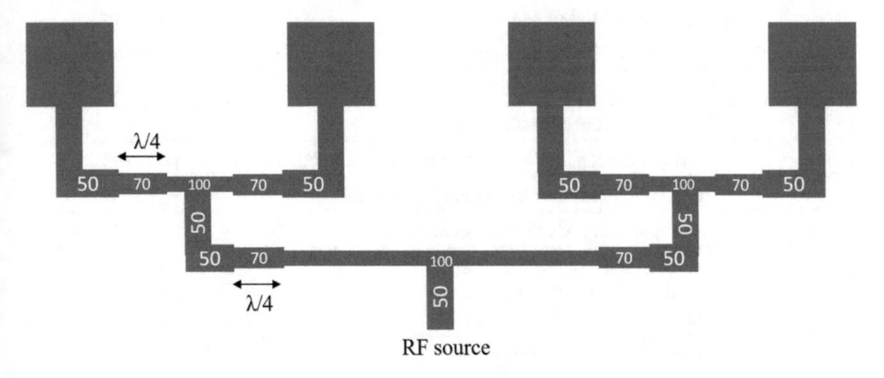

Fig. 4.10 Corporate feed design

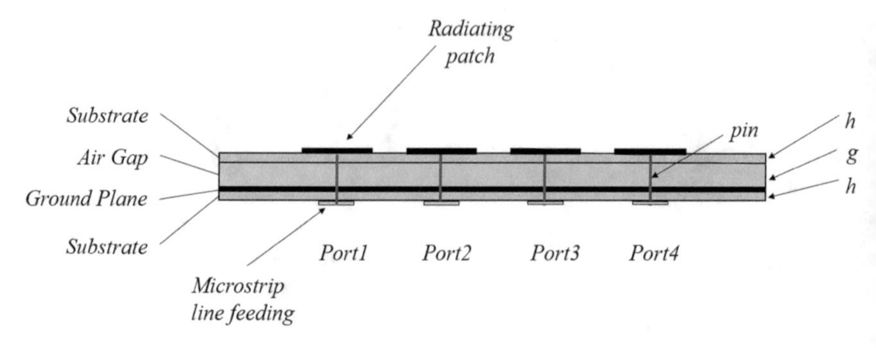

Fig. 4.11 Microstrip feed on the n × 4 antenna array

To address the requirement of the pre-configurable antenna system, we have created three variances of arrays using the same MPA element but with the reduced number of elements for lower gain version, i.e. 1 × 4 array (with 4 elements), 2 × 4 array (with 8 elements) and 4 × 4 array (with 16 elements). The dimensions of the respective array parameters are shown in Table 4.3. Except for the length of the arrays the rest of the design parameters are common to all the arrays. The common width for the arrays is 141.75 mm while the length of the arrays depends on the type of configuration, the smallest array with 1 × 4 configuration measured 60 mm in length, 2 × 4 array will take up 100.875 mm length and the biggest 4 × 4 array occupied 182.625 mm length.

The bandwidth ratio of the MPA has been improved using capacitive fee dual substrate technique, the bandwidth ratio improved to 43% compared to 4–7% for

Table 4.3 Design dimensions of the n × 4 antenna arrays

Parameter	Description	Value
h	The thickness of the die-electric substrate	1.56 mm
g	The distance of the air gap	6 mm
L	The length of the radiating element	15.5 mm
W	The width of the radiating element	16.4 mm
t	The length of the capacitive feed	3.5 mm
s	The width of the capacitive feed	1.7 mm
d	Air gap between the capacitive feed and radiating element	0.5 mm
y	Horizontal spacing of the antenna elements	27.25 mm
x	Vertical spacing of the antenna elements	40.875 mm
Lant	Length of the 1 × 4 antenna array Length of the 2 × 4 antenna array Length of the 4 × 4 antenna array	60 mm 100.875 mm 182.625 mm
Want	Width of the antenna array	141.75 mm
ε_r	Relative permittivity of the die-electric substrate	3

a typical MPA without any bandwidth enhancement. The HPBW, SLL and gain performance was optimised by iterating the element's spacing in the array, which is a practical simulation approach to determine the trade-off between the array's performance parameters which is important as it optimised the performance that suits the application, in our case, we are trying to achieve maximum gain while keeping the HPBW and SLL low. The corporate fed structure adopted can also simplify the design of the 1×4, 2×4 and 4×4 arrays that can be installed into the smart antenna systems that required different gain based on the field scenario as described in Figs. 3.2 and 3.3.

References

1. C.A. Balanis, Arrays and feed networks, in *Antenna Theory Analysis and Design*, 4th edn. (Wiley, Hoboken, New Jersey, USA, 2016), Chap. 14, Sect. 14.8, pp. 832–837
2. V.G. Kasabegoudar, D.S. Upadhyay, K.J. Vinoy, Design studies of ultra-wideband microstrip antennas with a small capacitive feed. Int. J. Antennas Propag. (2007). https://doi.org/10.1155/2007/67503
3. C.A. Balanis, Microstrip and mobile communications antennas, in *Antenna Theory Analysis and Design*, 4th edn. (Wiley, Hoboken, New Jersey, USA, 2016), Chap. 14, Sect. 14.2, pp. 788–823
4. 3ds.com, CST Studio Suite 3D EM Simulation and Analysis Software (2019), https://www.3ds.com/products-services/simulia/products/cst-studio-suite/. Accessed 29 May 2020
5. M.C. Tan, M. Li, Q.H. Abbasi, M. Imran, A wideband beam forming antenna array for 802.11ac and 4.9 GHz, in *2019 13th European Conference on Antennas and Propagation (EuCAP)*, Krakow, Poland, 2019, pp. 1–5

Chapter 5
Antenna Array Prototyping and Experimental Evaluation

The antennas were fabricated using the commercially available PCB fabrication technique. The PCBs that were fabricated and delivered from the factory are in panel form as demonstrated in Fig. 5.1, each array was separated by a V-cut for easy breaking before assembling into the respective antenna array. RO3003 material [1–3] from Roger (Arizona, United States) is the popular materials used in the high-frequency PCB industry, for cost reason, we have chosen a lowest cost alternative substrate F4BTM-2 produced by Taizhou Wangling (Jiangsu, China) with equivalent electrical specification compared to RO4003 materials, the new material is expected to pocket a 60% cost saving from the substrate material itself. The antenna arrays are constructed according to the stack-up structure shown in Fig. 4.11.

5.1 Assembling the Prototype Antenna Array

The construction of the 4 × 4 array can be presented in Fig. 5.2, (a) represents the stack-up view and (b) illustrates the exploded view. The top and bottom PCBs are supported by the metal feeding pins and 4 pieces of nylon spacers, the feeding pins are installed to transfer the RF energy from the feeding networks from bottom PCB to the capacitive feeding pads on the top PCB, the feeding pins are secured to both PCBs by solder. 2 nylon spacers (for 1 × 4 array) and 4 nylon spacer (for 2 × 4 and 4 × 4 arrays) are installed at the edge of the array to provide additional support to the PCBs as well as to maintain the air gap between the radiating PCB and the feeding PCB according to the design airgap of 6 mm, the nylon spacers are secured by the nylon nuts and screws. Nylon spacers, nylon screws and nylon nuts are chosen due to its excellent mechanical, thermal, and chemical properties and minimum affect to the electromagnetic performance of the antenna. The similar construction method is applied to 1 × 4 and 2 × 4 arrays but with a different number of elements.

M. C. Tan et al., *Antenna Design Challenges and Future Directions for Modern Transportation Market*, SpringerBriefs in Applied Sciences and Technology,
https://doi.org/10.1007/978-3-030-61581-9_5

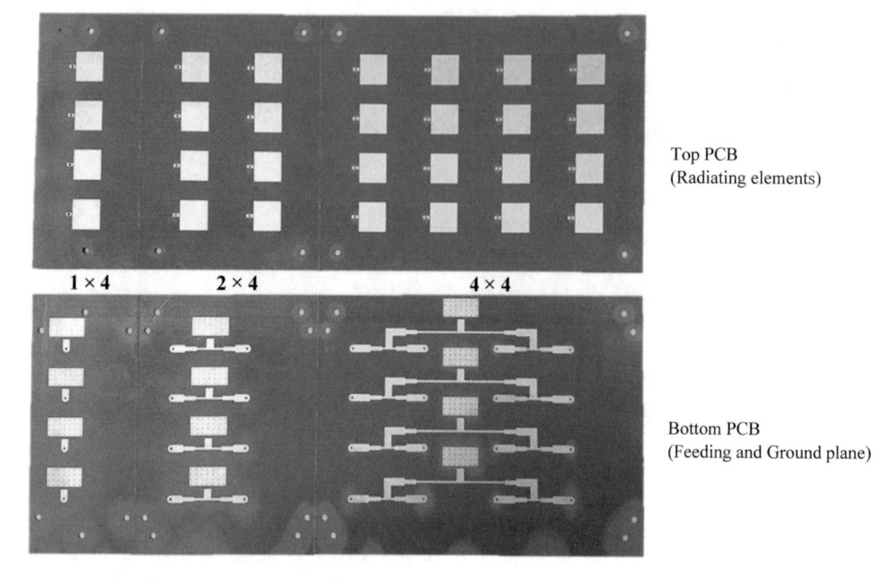

Fig. 5.1 Raw PCBs from the factory

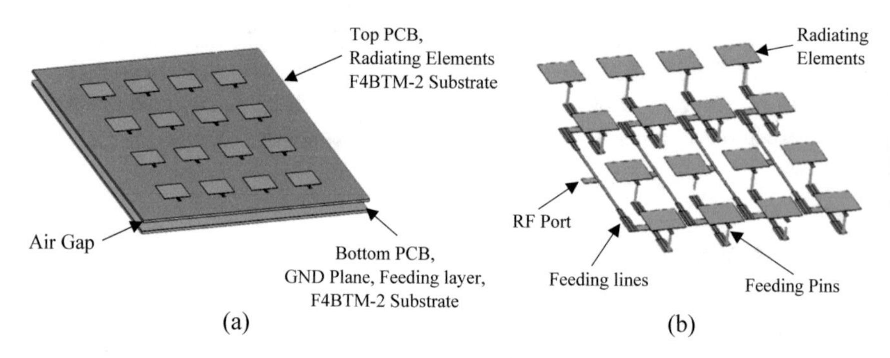

Fig. 5.2 The geometrical view of the 4 × 4 antenna array. **a** Stack up view. **b** Exploded view

To make sure we get the best performance out of the array prototype, some precautions are worth considering during the assembly process of the antenna array.

(i) PCB warpage, as the PCB area especially for the 4 × 4 array is bigger around 182.625 mm length and 141.75 mm width, the PCBs may potentially warp if the manufacturing control and packaging were not well-taken care by the PCB manufacturing factory. We have included 4 pieces of 6 mm nylon stand-off at the 4 corners of the PCBs, this is to ensure the top and bottom PCB are spaced evenly over the whole surface area. In mass production, there might be some visible warpage due to PCB manufacturing tolerance, a few more nylon washer

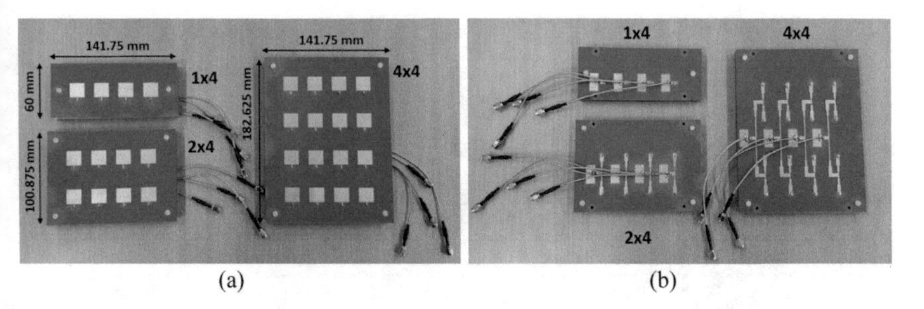

Fig. 5.3 Prototype photos of the arrays. **a** Top side. **b** Bottom side

can be installed at the middle or the side of the PCBs to properly hold the PCBs and offset the warpage.

(ii) The feeding pins and coaxial cables are soldered directly onto the PCBs, the soldering process must be properly controlled and ensure the good solder fillet and no dry joint at the solder joint. No excessive solder, as it may impact the characteristic impedance and matching of the microstrip line.

(iii) Ensure the RF coaxial at each feeding port are with equal length, this is to minimize the unnecessary phase variation due to unequal coaxial cable length at each port.

The coaxial cables are assembled to the antenna array by soldering the coaxial hot pin to the feeding copper pad and coaxial shield to the ground pad which was specially designed with sufficient copper opening for soldering purpose as demonstrated in Fig. 5.3b, The other end of the coaxial cables are assembled with the SMA connector that can be connected to test equipment or RF frontend directly. The final prototype of the antenna arrays are shown in Fig. 5.3, with both top and bottom views. Each antenna array comes with 4 mechanical mounting holes to facilitate the assembly process, where multiple antenna arrays will be mounted onto the radome to form a smart antenna structure that is designed to cover up to 360° scanning angle by combining multiple arrays with 90° coverage each.

5.2 Conducted Measurement

The returned loss and isolation between ports are measured in the conducted setup. The vector network analyser (VNA), model: VNA0406 e-SB manufactured by MegiQ (Eindhoven, Netherlands) with operating frequency from 400 MHz to 6 GHz is used to perform the measurement. The measured results of the S11 returned loss for the 3 types of proposed antenna arrays are presented in Fig. 5.4, all the antennas exhibited more than 10 dB returned loss for the intended band from 4.9 to 5.9 GHz, which is reasonable and within the antenna design requirement. A slight shift in the 1st resonance frequency point was observed around 4.5 GHz, which might be

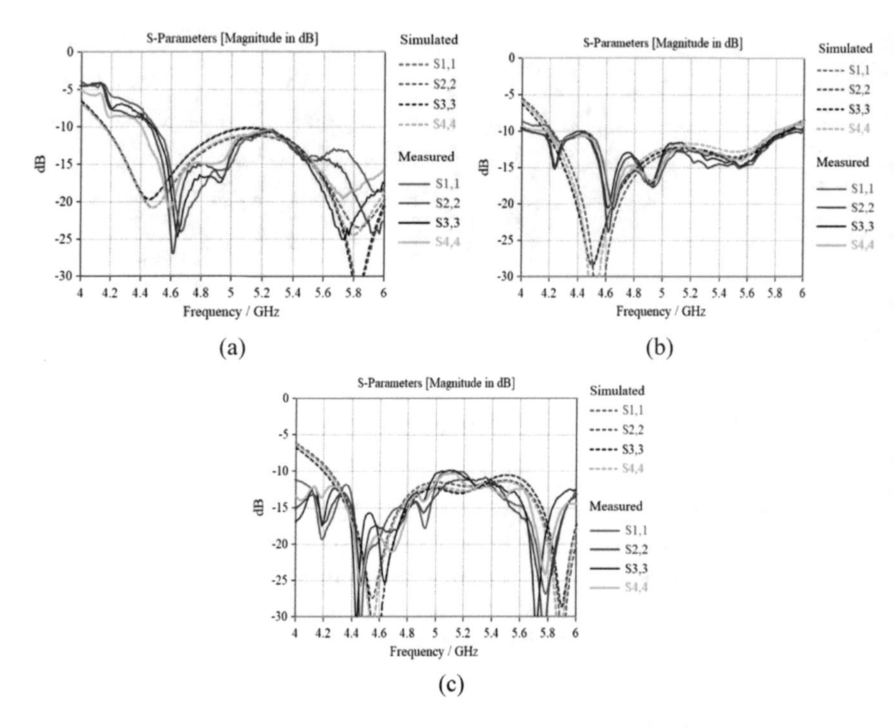

Fig. 5.4 Returned loss measurement results. **a** 1 × 4 antenna array. **b** 2 × 4 antenna array. **c** 4 × 4 antenna array

contributed by the manufacturing tolerance of the PCB, SMA connectors as well as the assembly tolerance. However, the overall design was able to fulfil the minimum 10 dB returned loss requirement set in our antenna design target over the 4.9–5.9 GHz band.

The port to port isolation was experimentally measured, for simplicity, due to the symmetric property of the antenna array, the inter port isolation measurement was conducted between the first port and the neighbouring ports, and the rest of the ports are expected to have the similar characteristic. Figure 5.5 shows the measurement results of the port to port isolation, a moderate agreement was demonstrated when compared with the simulated results, components tolerance, PCB manufacturing tolerance and stack up tolerance during the assembly of the antenna array might have contributed to the slight variation in the measured results. However, with the fair port to port isolation of minimum 15 dB over the 4.9–5.9 GHz band is acceptable to eliminate the crosstalk that may potentially affects the performance of the array.

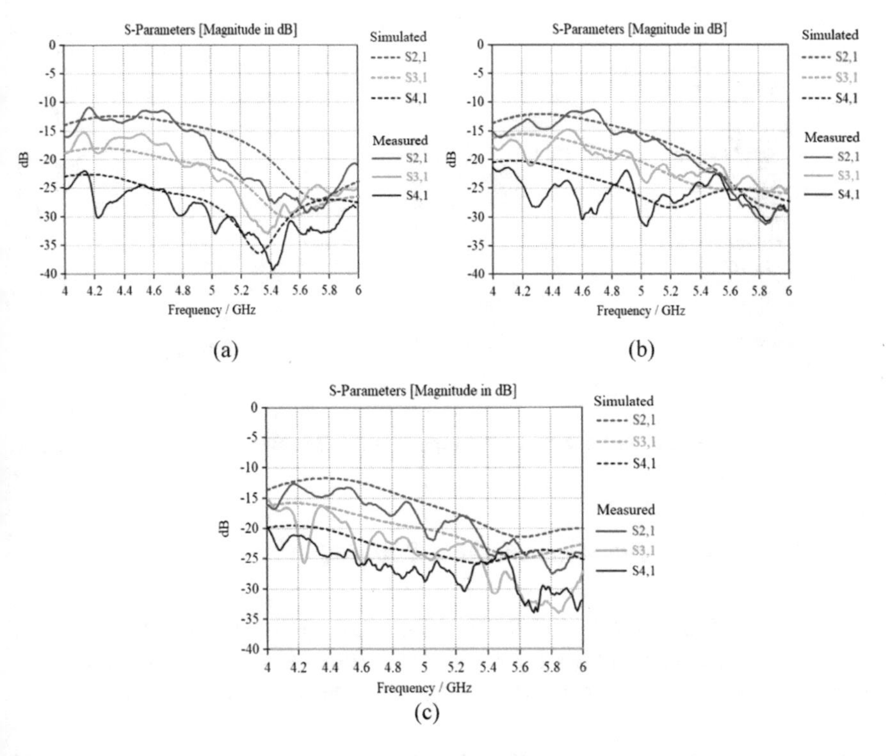

Fig. 5.5 Simulation and measured Inter-port isolation results. **a** 1 × 4 array. **b** 2 × 4 array. **c** 4 × 4 array

5.3 Radiated Measurement for the Antenna Arrays

To ease the iteration process, we have set up the open site to perform the far-field measurement of the antenna radiation pattern including the gain, sidelobe level and beam steering performance, the measurement setup is demonstrated in Fig. 5.6, a receiving antenna with the known gain was placed at a distance *D* away from the antenna under test, the radiated power transmitted from the antenna under test is captured by the receiving antenna and feed into the spectrum analyser for measurement and data logging. This section will cover the far-field antenna radiation pattern measurement at its nominal 0° beam. A power splitter is used to split the RF source into 4 dedicated antenna ports that connected to the antenna under test. For 0° beam measurement, the 4 antenna ports are injected with the signal with equal phase.

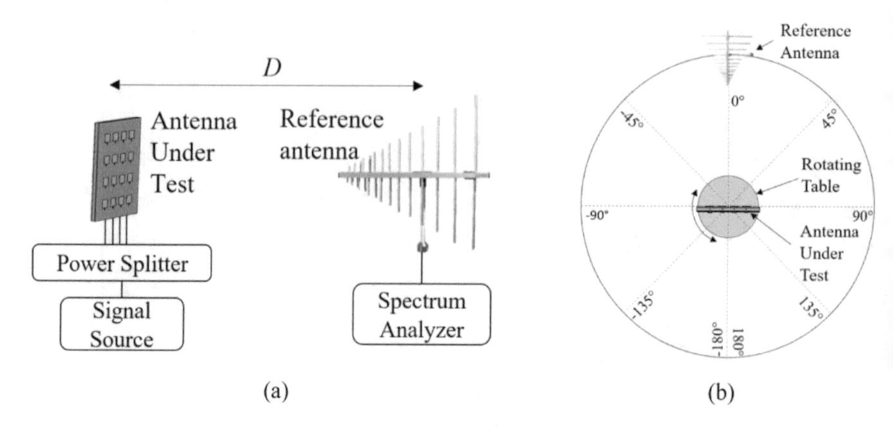

Fig. 5.6 Antenna measurement setup in open space. **a** Antenna measurement setup. **b** Antenna orientation (top view)

5.3.1 *The Antenna Measurement Topology*

Open site technique [4] along with the Free Space Loss (FSL) method [5] was adopted to measure the antenna radiation pattern of the prototype arrays. Thus, the antenna gain of the receiver can be calculated using Friis equation as shown below:

$$G_r + G_t = 20\log\left(\frac{4\pi f D}{c}\right) + P_r - P_t + \alpha \tag{5.1}$$

where G_r and G_t are referring to the realized gain in (dBi) of the golden antenna (receiver antenna) and the target antenna (transmitter antenna) respectively, the interest frequency is represented by f, Distance between the transmitter antenna and receiver antenna is represented by D, $c = 3 \times 10^8$ m/s denotes the speed of light, P_r is the power received by the reference antenna in (dBm), P_t represents the power transmitted by the signal source in (dBm), the attenuator factor due to connectors and cables is shown as α.

In the measurement setup, the log-periodic antenna part number LP-04 is used as a reference antenna, the antenna is manufactured by Narda (Milano, Italy) with the gain provided by the manufacturer is 6 dBi at 5.5 GHz. The spectrum analyser E4407B from Agilent (Santa Clara, United States) is used to measure the power received by the reference antenna. A pre-calibrated off the shelf 802.11ac wireless dongle part number WUSB-3002 is used as signal source to provide the transmit output power of 23 dBm at the frequencies between 5.1 and 5.9 GHz. In the measurement setup with D = 3, the calculated FSL is 56.79 dB, the measurement system was pre-calibrated to include the offset of the FSL and α, attenuation factor.

To simulate the radiating angle during the measurement, the antenna under test is placed on the turntable as shown in Fig. 5.6b and constantly rotates through the azimuth plan in the step of 5° over the 360° sector to simulated the 360° radiation pattern. The receiver antenna will pick up the RF field while the transmitter antenna

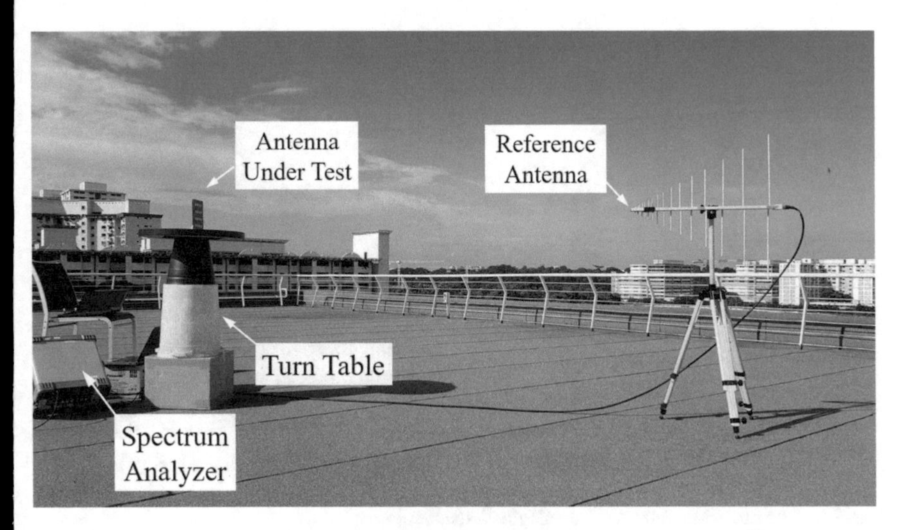

Fig. 5.7 Photo of the far-field antenna measurement setup

rotates, the measured power can be converted to the antenna gain using Eq. (5.1) and presented as antenna radiation pattern by plotting the full set of measurement data over the 360° chart. The photo of the measurement setup is shown in Fig. 5.7, the measurement was done at the open site at the building top to minimise the measurement uncertainly due to multipath RF reflection.

5.3.2 Radiation Measurement of the Single Element Antenna

The single element was experimentally evaluated using the measurement setup in Fig. 5.7, the RF beamwidth observed is around 80° and matched well with the simulated result as shown in Fig. 5.8. The far-field antenna gain is measured at 3 frequencies 4.9, 5.5 and 5.9 GHz with the gain of 7.3, 687 and 5.5 dBi respectively, compared to the simulated results of 7.42, 8.59 and 8.69 dBi, there is a variation of gain at the high band, this might be due to the frequency response of the coaxial cable and RF connectors.

5.3.3 The Radiation Pattern of the Antenna Arrays

The boresight antenna gain at 0° and 5.5 GHz of the 1×4, 2×4 and 4×4 antenna arrays are shown in Fig. 5.9, and Table 5.1 tabulated the key antenna performance parameter such as the far-field gain, HPBW, SLL, with the comparison between both measured and simulated results.

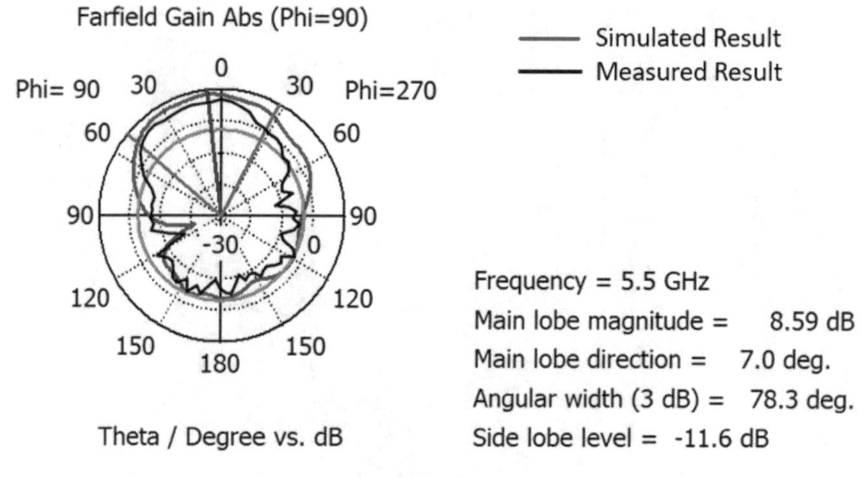

Fig. 5.8 The radiation pattern of the single element antenna

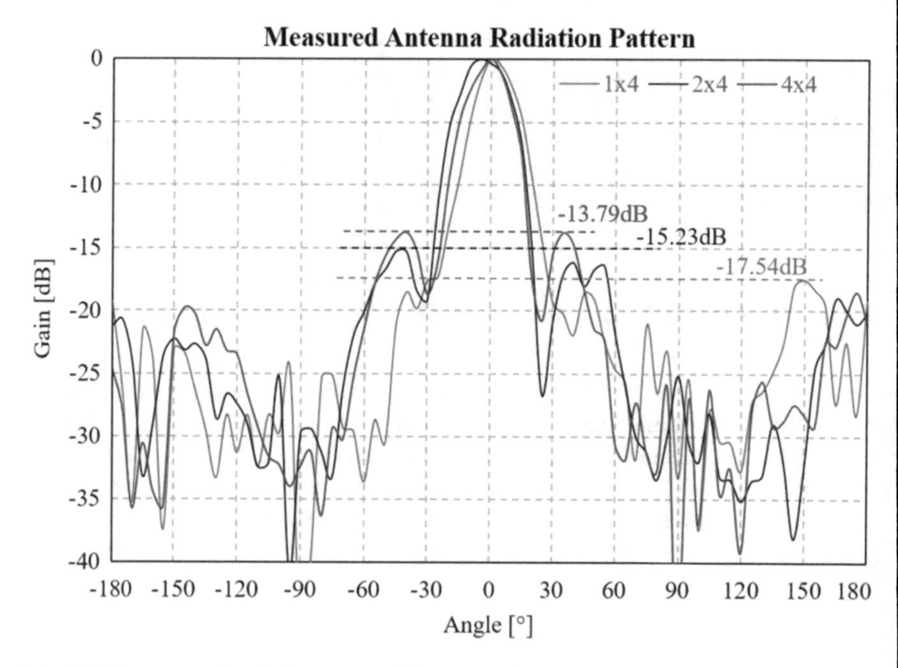

Fig. 5.9 The measured radiation pattern of the proposed antenna arrays

Referring to the results in Fig. 5.9 and Table 5.1, the measured antenna gains for 1 × 4, 2 × 4 and 4 × 4 arrays are 11.16, 14.59 and 17.25 dBi respectively, which is tallied with the simulated results of 13.53, 15.4 and 17.8 dBi, some slight deviation that might be caused by the tolerance of PCB manufacturing, coaxial cable, connectors, as well as the environmental factors. Further evaluation on the realized

Table 5.1 Evaluation results of the 1 × 4, 2 × 4, and 4 × 4 antenna arrays

			1 × 4	2 × 4	4 × 4
Antenna gain	dBi	Simulated	13.53	15.4	17.8
		Measured	11.16	14.59	17.25
Sidelobe level	dB	Simulated	−12.9	−13.8	−13.6
		Measured	−17.54	−15.23	−13.79
3 dB beamwidth	°	Simulated	24.4°	24.3°	24.5°
		Measured	20°	24°	20°

gain was carried out for the entire interest band from 4.9 to 5.9 GHz for the low, middle and high gain arrays, their respective gains are 11.16–12.27 dBi, 13.97–15.74 dBi and 17.25–18.12 dBi, the gain flatness over the interest frequency band is kept within 2 dB which is reasonably good for the wideband antenna. The results show a moderate agreement with the simulated result of 12.90–14.10 dBi, 15.00–15.70 dBi and 17.25–18.12 dBi for the respective low, middle, and high gain arrays. There is a slight improvement of 4.5 dB for the SLL as compared to the simulation results, the measure SSL result for the low gain array is −17.54 dB compared to −12.9 dB of the simulated result, the main reason was due to the slight lower gain of the antenna prototype that leads to the improvement in the SSL level. In the case of HPBW, the measured result of less than 24° which is also tallied well with the simulated result with the SLL value of around 24°. The proposed middle gain 2 × 4 antenna (with 8 elements) is compared with the state-of-the-art 8-elements antenna array in [6], the gain of the proposed antenna outperformed the design in [6] with the gain at 0° of 13.97–15.74 dBi compared with [6] with just 6.9 dBi, however, both arrays demonstrated less than −10 dB SLL. Therefore, we conclude that the proposed low, middle and high gain array are the suitable candidate that can be integrated into the beamforming antenna system which requires a superior gain and SLL performance.

References

1. C.A. Balanis, Microstrip and mobile communications antennas, in *Antenna Theory Analysis and Design*, 4th edn. (Wiley, Hoboken, New Jersey, USA, 2016), Chap. 14, Sect. 14.2, pp. 788–823
2. M. Mantash, T.A. Denidni, Millimeter-wave beam-steering antenna array for 5G applications, in *IEEE 28th Annual International Symposium on Personal, Indoor, and Mobile Radio Communications (PIMRC)*, Montreal, QC, Canada, 2017, pp. 1–3
3. V.G. Kasabegoudar, D.S. Upadhyay, K.J. Vinoy, Design studies of ultra-wideband microstrip antennas with a small capacitive feed. Int. J. Antennas Propag. (2007). https://doi.org/10.1155/2007/67503
4. S.D. Assimonis, T. Samaras, V. Fusco, Analysis of the microstrip-grid array antenna and proposal of a new high-gain, low-complexity and planar long-range WiFi antenna. IET Microwaves Antennas Propag. **12**(3), 332–338 (2018). https://doi.org/10.1049/iet-map.2017.0548

5. A. Ghasemi, A. Abedi, F. Ghasemi, Introduction to radiowaves propagation, in *Propagation Engineering in Radio Links Design* (Springer, New York, USA, 2013), Chap. 1, Sect. 1.10.1, pp. 26–27
6. Y. Wang, L. Zhu, H. Wang, Y. Luo, G. Yang, A compact, scanning tightly coupled dipole array with parasitic strips for next-generation wireless applications. IEEE Antennas Wireless Propag. Lett. **17**(4), 534–537 (2018). https://doi.org/10.1109/LAWP.2018.2798660

Chapter 6
Evaluation of the 360° Antenna Systems

6.1 Beamforming Performance Evaluation

The beamforming performances of the 1×4, 2×4 and 4×4 arrays will be covered in this section, including the results for both simulation and experimental measurement, the simulation was done using the CST simulation tools and the measurements are carried out using the measurement topology explained earlier. To the end of this section, we have included the construction method for the 360° beamforming antenna structure and presented a table to highlight the advantages of the proposed pre-configurable antenna structure over the related state-of-the-art antennas that have the same 360° steering coverage.

6.2 Simulation Results of the Beamforming Antenna Array

CST tool is used to simulate and validate the beamforming performance of the antenna arrays, the phase of each antenna port from P1 to P4 are varied so that the beam direction of the antenna can be controlled. The phase shift of the individual antenna feeding ports from P1 to P4 with respect to the beam direction can be explained as follows, the phase shift of 0°/0°/0°/0° for 0° beam direction, 45°/180°/−45°/90° for −40° beam direction and 90°/−45°/180°/45° for +40° beam direction. The simulated results of the beamforming performance of the 1×4, 2×4 and 4×4 arrays can be found in Fig. 6.1, all the arrays demonstrated a ±41° beam steering capability, when operating at HPBW, the steering angle shall cover up to ±45°.

M. C. Tan et al., *Antenna Design Challenges and Future Directions for Modern Transportation Market*, SpringerBriefs in Applied Sciences and Technology,
https://doi.org/10.1007/978-3-030-61581-9_6

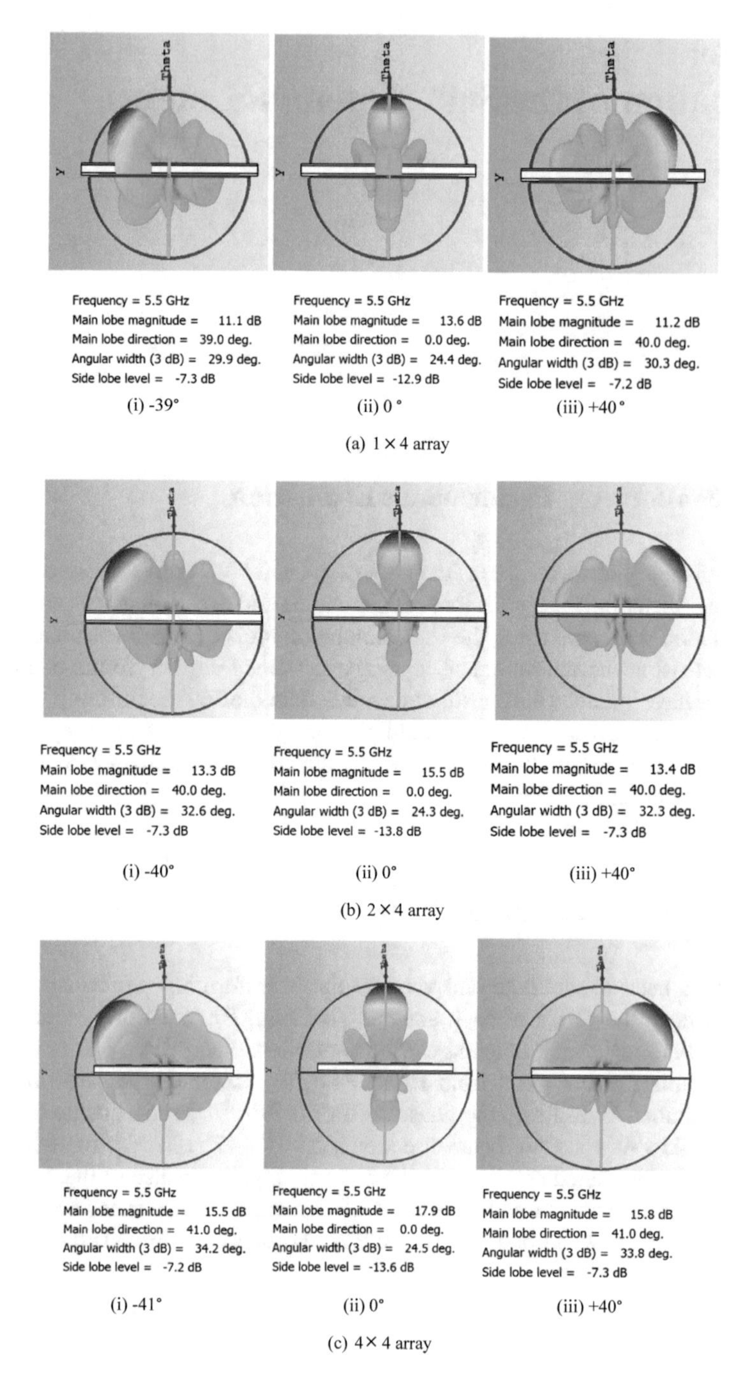

Fig. 6.1 Simulated beamforming performance of the 1×4, 2×4 and 4×4 arrays

6.3 Experimental Verification on the 360° Beamforming Systems

The beam steering performance of the proposed 4 × 4 array is further evaluated using the flexible beamforming concept in [1] and the setup in Fig. 6.2. The flexible structure was constructed by integrating with maximum 4 sets of 90° phased arrays indicated as A1, A2, A3 and A4 to support 90°, 180°, 270° and the entire 360° service sector, each of the phased array is connected via 4 horizontal feeding ports C1, C2, C3 and C4 to support the beamforming in the dedicated 90° sector. For 360° coverage, a total of 4 sets of arrays with 16 feeding ports will be connected to the 4 RF beamforming chains. The key building blocks of the RF chain are explained below,

i. A Single Pole Four Throw (SP4T) switch for RF switching between the antenna column,
ii. Two Single Pole Double Throw (SPDT) switches to switch the RF path between the transmitter and receiver,
iii. A 6-bit digital phase shifter to control the phase of each antenna column,
iv. A Low Noise Amplifier (LNA) for the receiver to boost the signal received from the antenna and,
v. A Power Amplifier (PA) for power amplification during transmission,
vi. A RF Attenuator (ATT) with 30 dB attenuation range to control the transmit power level.

Another function of the LNA and PA is to compensate for the insertion loss exhibited by onboard components such as SPDT, SP4T, phase shifter and power combiner.

A power combiner is used to channel the RF signal from the 4 RF chains before it is connected to the radio transceiver and baseband processor, the antennas switching and the selection of transmitter/receiver mode are controlled by external processor unit such as microcontroller or Field Programmable Gate Array (FPGA) via various interfaces such as Input–Output (IO) signals, Digital to Analog Converter (DAC)

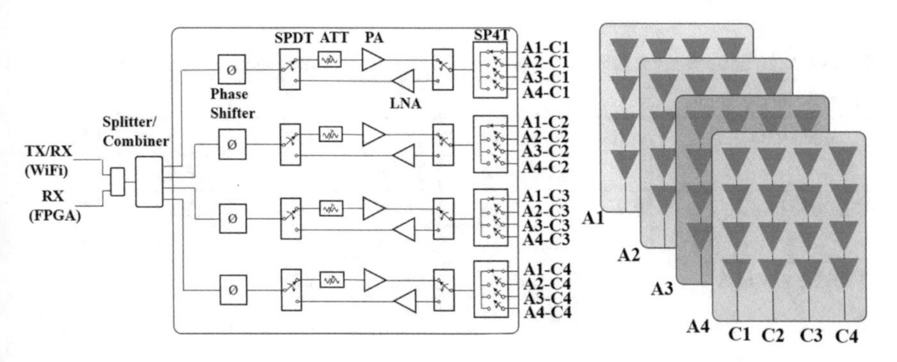

Fig. 6.2 Overview of the 360° pre-configurable antenna structure

and Analog to Digital Converter (ADC) that directly interacts, manage and control the respective block within the RF chains.

To simplify the beamforming evaluation process, the evaluation was done on one of the 90° sectors and due to the similarity of the RF chains and arrays, the rest of the sectors are expected to have the similar beamforming performance. In this measurement, one of the 4 arrays in the beamforming antenna system is chosen, for instance, A4 was chosen in this case, and the 4 RF chains are toggled to switch to the respective antenna column in A4 via the SP4T switch. The measurement setup and test environment are similar to Figs. 5.6 and 5.7. The beamforming result of the 4 × 4 arrays is presented in Fig. 6.3.

Table 6.1 illustrates the simulated and measured beamforming performance of the 4 × 4 array. A fair agreement was observed between the two results. Since all the arrays and RF chains are identical, similar beamforming result is expected for the other arrays that cover the entire 360°. The simulated and measured beam scanning angle resolution of the proposed arrays was approximately 2°, this was achieved by using the phase shifter with lowest phase resolution of 5.625°.

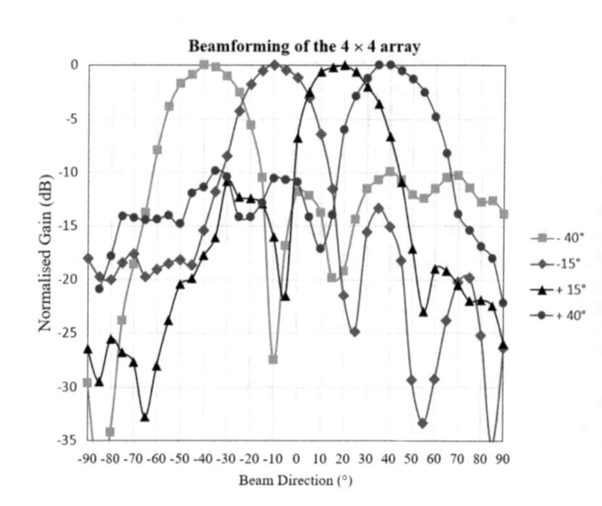

Fig. 6.3 Measured beamforming results of the 4 × 4 array

Table 6.1 Beamforming results of the proposed 4 × 4 array (simulated and measured)

Beam steering angle			−40°	−15°	15°	−40°
Antenna gain	dBi	Simulated	15.8	17.6	17.5	15.5
		Measured	14.38	16.03	16.43	14.79
Sidelobe level	dB	Simulated	−7.3	−12.0	−11.6	−7.2
		Measured	−9.95	−13.34	−12.25	−9.84
3 dB Beamwidth	°	Simulated	33.8	25.9	25.9	34.2
		Measured	30.9	27.5	28.2	31.5

6.4 Construction of the 360° Beamforming System

The 360° beamforming antenna system consists of 4 sets of antenna arrays with each covering 90°, the beamforming is realised using the RF beamforming frontend also known as transmit and receive (T/R) module as presented in [2. 3], The Papers have highlighted some key design requirement for the T/R modules and necessary calibration technique required to offset the possible phase and amplitude error during manufacturing, the technique to reduce the SLL of the beamforming antenna system has been discussed in detail. In typical deployment in the transportation sector, the smart antenna will be deployed outdoor or installed on the vehicle, therefore, the antenna arrays and electronics need to be properly housed inside the enclosure or radome to protect it from the harsh environment, in [4] a study was done to assess the potential RF degradation due to the antenna radome as well as the effect of the drag coefficient to the shape of the radome, the results revealed that the polymethyl methacrylate material has negligible impact to the radiation performance at 4.9–5.9 GHz, and demonstrated that the improved aerodynamic design of the radome has a minimum effect to the overall drag coefficient of the vehicle.

The beamforming antenna systems will be pre-configured according to the application scenarios described earlier in Fig. 2.5, the proposed flexible architecture allows a maximum of 4 sets of arrays to be integrated and form a smart antenna system. The 360° beamforming structure is constructed using the 4 units of 4 × 4 arrays as demonstrated in Fig. 6.4 with both the isometric view and 360° radiation pattern simulated using CST software. Overall antenna performance was measured with the gain between 14.38–17.25 dBi with 20°–31.5° radiating beamwidth.

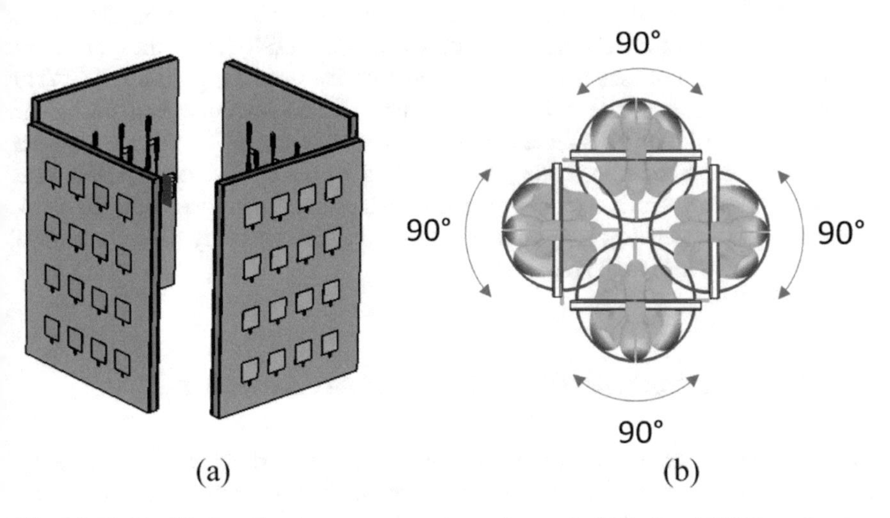

Fig. 6.4 The flexible beamforming antenna structure. **a** Isometric CAD view. **b** 360° beamforming pattern

Each of the 90° sectors is designed to support (i) full 90° scanning with only 1 antenna port is enabled, or (ii) beam steering with 25° narrow beamwidth by enabling all the ports with phase control on each port. The features are very useful for direction of arrival (DOA) estimation in the 360° smart antenna system as presented in [1], where the sector scanning to determine the specific sector of the target, this can be performed by enabling one of the 4 columns on the 4 sector arrays, once the target is locked to the targeted sector, the fine direction of arrival can be performed by enabling all the 4 columns on that particular array and zoom in the beam to the target with a narrow beamwidth of approximately 25°, beam scanning with 25° beamwidth and scanning resolution of up to 2° within the 90° sector is possible with the 6-bits digital phase shifter that is capable of producing the phase shift resolution as low as 5.625° on each RF chain.

The RF beamforming frontend is controlled by the external processor such as FPGA or single-board computer, the function will be briefly discussed in this section, the baseband processing consists of the Direction Of Arrival (DOA) algorithm combined with the sophisticated beamforming algorithm to compute the necessary beamforming weight that controls the hybrid beamforming frontend to achieve the desired beamforming and interference nulling. Various works have been carried out to optimise the DOA algorithm, such as the DOA performance evaluation [5] on the 4-elements array using Root-Multiple Signal Classification (Root-MUSIC), Root-Weighted Subspace Fitting (Root-WSF) and Beamspace-Estimation of Signal Parameters via Rational Invariance Techniques (BSESPRIT) algorithms, performance optimization such as reducing the DOA computational time [6] via the Root-Transformation Matrix (root-T) technique, for low Signal to Noise Ratio (SNR) environment scenario, the Cross Cumulant-MUSIC (CC-MUSIC) combined with the root-MUSIC algorithm [7] was proposed. In the area of beamforming algorithm, the blocking matrix for Generalized Sidelobe Canceller (GSC) [8] for the high-speed environment was proposed using a simplified Zero Placement Algorithm (SZPA) to achieve reduced computational time to 0.268 ms from 1.541 ms compared to Singular Value Decomposition (SVD) method. In [9], the improvement on the convergence speed in the adaptive beamforming was proposed via the Cyclic Variable Step Size (CVSS) algorithm. The Kalman Filter was introduced in [10] to overcome the DOA angle mismatch when the changes of the DOA is too fast especially when the mobile terminal is very close to the receiver station.

6.5 Comparison with the State-of-the-Art Antennas

This section will summarize the pre-configurable antenna performance by comparing the key performance parameters with the similar 360° steering capability in the 802.11ac application field taking into consideration the application scenarios as elaborated earlier in Fig. 2.5. The comparison result can be found in Table 6.2. The advantages of the proposed pre-configurable antenna system over the state-of-the-art antenna systems are highlighted below,

Table 6.2 Comparison between the proposed antenna system and the state-of-the-art beamforming antenna

	[11]	[12]	This work
Type of element in the array	Electro-magnetic bandgap (EBG)	Circular patch + six parasitic elements	Dual-substrate capacitive-feed MPA
Number of arrays	12	6	4
Operating frequency (GHz)	5.18–5.825	5.1–5.9	4.9–5.9
Antenna array gain (dBi)	12	10	14.38–17.25
Azimuth beamwidth	30°	42°	27.5°–31.5°
Beam steering method	PIN diode	PIN diode	SP4T and phase shifter
Beam steering resolution	30°	60°	Minimum 2°
SLL (dB) (read from the antenna radiation graph)	−6 to −12	−7	−9.8 to −13.3

(i) **Wide operating band**, is realized by multiple substrate structure that enables the wide operating bandwidth, the proposed design covers the entire 4.9–5.9 GHz which include the 5 GHz ISM band and licensed band at 4.9 GHz.
(ii) **Novel array structure**, that radiated power are combined and concentrated in a small beamwidth that effectively improved the antenna gain to 14.38–17.25 dBi compared to 10 and 12 dBi, that makes it more suitable for long-range application such as transportation environment.
(iii) **Flexible beam steering capability**, dedicated phase shifter for each RF chain, and each sector is powered by 4 units of RF chains with the phase of each chain can be independently set.
(iv) **Pre-configurable gain**, the gain of the antenna system can be pre-configured before it is installed, where the low, middle and high gain arrays with different gain can be pre-selected and built into the 360° beamforming system.
(v) **Pre-configurable sector**, the flexible structure can be pre-configured to 90°, 180°, 270° and 360° coverage with each sector cover 90°.
(vi) **Fine beam steering angle**, the 6-bit digital phase shifter with 5.625° phase resolution enable the lowest beam resolution of 2°. The beamforming accuracy is improved with the fine beam steering resolution.

References

1. M.C. Tan, M. Li, Q.H. Abbasi, M. Imran, A flexible low-cost hybrid beamforming structure for practical beamforming applications, in *2019 IEEE International Symposium on Radio-Frequency Integration Technology (RFIT)*, Nanjing, China, 2019, pp. 1–3. https://doi.org/10.1109/RFIT.2019.8929162

2. M.C. Tan, M. Li, Q.H. Abbasi, M. Imran, A recursive calibration approach for smart antenna beamforming frontend, in *2020 14th European Conference on Antennas and Propagation (EuCAP), Copenhagen*, Denmark, 2020, pp. 1–5. https://doi.org/10.23919/EuCAP48036.2020.9135881
3. M.C. Tan, M. Li, Q.H. Abbasi, M.A. Imran, Design and characterization of T/R module for commercial beamforming applications. IEEE Access **8**, 130252–130262 (2020). https://doi.org/10.1109/ACCESS.2020.3009531
4. D.T.R. Liang, M.C. Tan, M. Li, Q.H. Abbasi, M. Imran, Radome design with improved aerodynamics and radiation for smart antennas in automotive applications, in *2019 IEEE International Symposium on Radio-Frequency Integration Technology (RFIT)*, Nanjing, China, 2019, pp. 1–3. https://doi.org/10.1109/RFIT.2019.8929217
5. M. Muhammad, M. Li, Q.H. Abbasi, C. Goh, M. Imran, Performance evaluation for direction of arrival estimation using 4-element linear array, in *2019 13th European Conference on Antennas and Propagation (EuCAP)*, Krakow, Poland, 2019, pp. 1–5
6. M. Muhammad, M. Li, Q.H. Abbasi, C. Goh, M. Imran, Direction of arrival estimation using root-transformation matrix technique, in *2019 IEEE International Symposium on Antennas and Propagation and USNC-URSI Radio Science Meeting*, Atlanta, GA, USA, 2019, pp. 1369–1370. https://doi.org/10.1109/APUSNCURSINRSM.2019.8889249
7. M. Muhammad, M. Li, Q.H. Abbasi, C. Goh, M. Imran, Direction of arrival estimation using hybrid spatial cross-cumulants and root-MUSIC, in *2020 14th European Conference on Antennas and Propagation (EuCAP)*, Copenhagen, Denmark, 2020, pp. 1–5. https://doi.org/10.23919/EuCAP48036.2020.9135813
8. S. Dai, M. Li, Q.H. Abbasi, M.A. Imran, A fast blocking matrix generating algorithm for generalized sidelobe canceller beamformer in high speed rail like scenario. IEEE Sens. J. https://doi.org/10.1109/JSEN.2020.3002699
9. S. Dai, M. Li, Q.H. Abbasi, M. Imran, Hardware efficient adaptive beamformer based on cyclic variable step size, in *2018 IEEE International Symposium on Antennas and Propagation and USNC/URSI National Radio Science Meeting*, Boston, MA, 2018, pp. 191–192. https://doi.org/10.1109/APUSNCURSINRSM.2018.8608636
10. S. Dai, Q.H. Abbasi, M. Li, M. Imran, Beamforming optimization based on Kalman filter for vehicle in constrained route, in *2019 IEEE International Symposium on Antennas and Propagation and USNC-URSI Radio Science Meeting*, Atlanta, GA, USA, 2019, pp. 1365–1366. https://doi.org/10.1109/APUSNCURSINRSM.2019.8888482
11. H. Boutayeb, P.R. Watson, W. Lu, T. Wu, Beam switching dual polarized antenna array with reconfigurable radial waveguide power dividers. IEEE Trans. Antennas Propag. **65**(4), 1807–1814 (2016). https://doi.org/10.1109/TAP.2016.2629469
12. Y. Yang, X. Zhu, A wideband reconfigurable antenna with 360° beam-steering for 802.11ac WLAN applications. IEEE Trans. Antennas Propag. **66**(2), 600–608 (2017). https://doi.org/10.1109/TAP.2017.2784438

Chapter 7
Conclusions and Future Directions

The advancement of IoT has enabled the various type of sensors onboard the vehicles, the most relevant to automotive sensors such as GPS, accelerometer, gyroscope or magnetometer which can be easily integrated into communication gateway or commercially available as off the shelf standalone module. Such sensors may already be available in many of the vehicles in the public transportation sector. With the aids of sensors, we believe the beamforming will no longer need to act alone.

Considering the operation of the public buses and train, the servicing routes for the buses are fixed according to their service map, similarly to train, where the rail tracks are fixed and the trains always follow the dedicated route when in operation. This is one of the advantages that the smart beamforming antenna designers can leverage on. With the fix geographical location of the wireless infrastructure being installed at the roadside or rail trackside, and the fix location of the equipment being mounted in the vehicles, as well as the predetermined direction and route of the vehicle on the move, we can combine the information gathered from the onboard sensors such as GPS, accelerometer, gyroscope, odometer, Bluetooth low energy (BLE) etc., utilizing the sensors fusion method by combining the sensory data with the native beamforming algorithm to produce a more accurate beamforming direction between the base stations and the mobile clients. This method is also set to reduce the complexity and processing effort of the native beamforming algorithm when combined with the sensor fusion.

7.1 Type of IoT Sensors Commonly Presents in the Vehicular Environment

GPS is the most common sensors in the vehicular industry, it received signal from multiple satellites to produce and determine the useful positioning data, utilising the simple and matured specification called National Marine Electronics Association,

M. C. Tan et al., *Antenna Design Challenges and Future Directions for Modern Transportation Market*, SpringerBriefs in Applied Sciences and Technology,
https://doi.org/10.1007/978-3-030-61581-9_7

NMEA-0183 to communicate with the host processor. The protocol provides the geographical position of the receiver such as longitude, latitude, and altitude, as well as time, speed and heading information. It is widely used for vehicle positioning and navigation-related applications. When the vehicle is moving into the underground tunnel for a short instance of time where it lost the GPS signal, the Dead Reckoning feature [1] can recover the navigation data with the aid of other sensors such as accelerometer, gyro meter and odometer.

The accelerometer detects the acceleration of the vehicle, the acceleration can be converted to velocity, micro-electro-mechanical systems (MEMS) accelerometer is the most used accelerometer in electronic devices. Gyroscope sensor is to measure the rotational angle of the moving vehicle, the deviation of the vehicle from its original orientation can be measured. The odometer is used to measure the distance travelled by the vehicle. Most of the vehicles already have some of the sensors integrated into the vehicles control system, the sensors information can be retrieved via the Controller Area Network (CAN bus) interface.

7.2 The Architecture of the Future Beamforming Antenna System

The vehicle is pre-loaded with map includes the geographical locations of all the access points installed as the roadside infrastructure, when the vehicle is on the move, the sensors data will be collected and through the aids of sensor fusion to produce the vehicle information such as location, speed, time and heading information, with the known location of the access points, the system will be able to determine which access point the vehicle should connect to and at what angle the beamforming antenna should concentrate its beam in order to have optimum point to point connections. When the vehicle is passed the access point, the system can predict which is the next access point and its location for the moving vehicle to roam to, while the mobile client prepared to roam over, the beamforming engine will dynamically steer its beam towards the targeted access point. The location and heading information of the moving vehicle will be transmitted to the backend, the back end will use this information to determine the location and direction of the mobile vehicle with respect to the access point that serves the connection and steers its beam toward to vehicle, similarly, when the mobile client passed the access point, the system is able to predict which access point is next in the queue to serve the approaching vehicle.

The sensors fusion diagram combined with the light direction of arrival (DOA) engine of the next generation beamforming system is presented in Fig. 7.1. With this hybrid approach by combining the sensor fusion output with the light computation output from the direction of arrival engine, it can further improve the accuracy of the beam steering compare to the method with DOA engine itself. Furthermore, the existence of sensor data will reduce the processing complexity of the DOA engine, hence reducing the processing power of the DOA module.

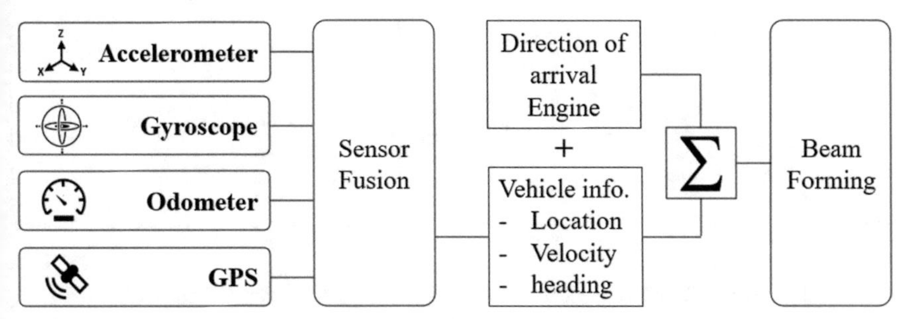

Fig. 7.1 Combination of sensor fusion and light direction of arrival for future beamforming antenna system

7.3 Benefits Gained from the Vehicular Operating Scenarios

The illustrations of the IoT sensors aided beamforming system is presented in Fig. 7.2. We would like to discuss 2 operation scenarios in this setup, (i) with reliable GPS data, for example, clear sky and (ii) when GPS signal is not available such as the vehicle is inside the tunnels or its moving in the urban area with high rise building where reliable GPS signal is limited.

For scenario (i) operation with reliable GPS data, the access points that are permanently mounted at the roadside are with known geographical coordinate and the mobile clients can obtain their location information from the on-board GPS receiver. The 2 access points are located at the location *a* and *b* as the roadside infrastructure to provide communication for the passing by vehicles, the vehicle with wireless mobile clients and GPS receiver installed is moving along the road heading to the right, the mobile client will bypass the access points a and b, and $m1$, $m2$ and $m3$ are the locations of the mobile client along the road. While moving, the wireless client in the vehicle will roam from AP_a to AP_b. The top view of this operation scenario can be represented as a diagram in Fig. 7.3. The symbols in the diagram can be explained as follow, the φ and θ are the bearings between the mobile client and the AP with

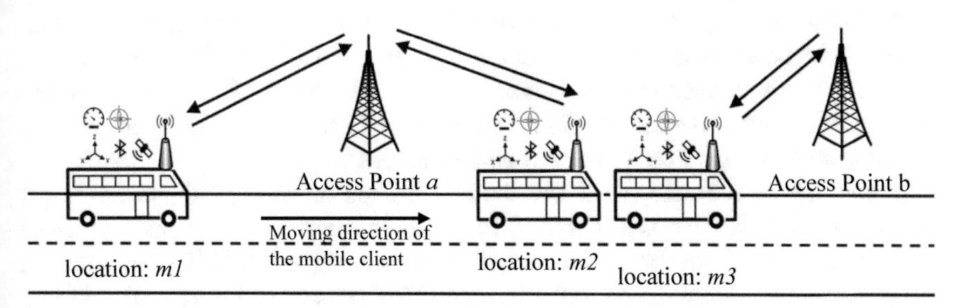

Fig. 7.2 IOT sensors aided beamforming system

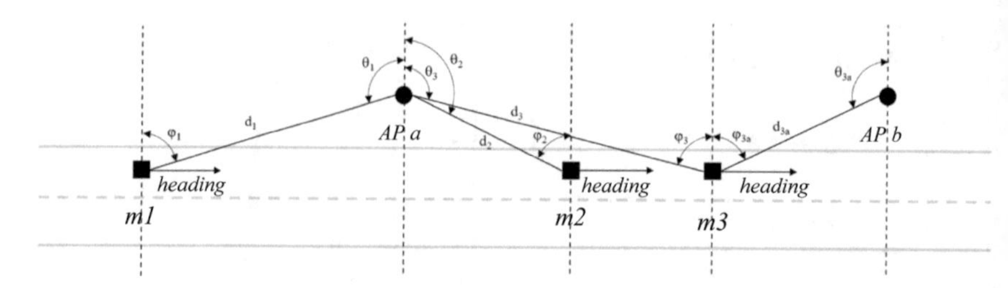

Fig. 7.3 Top view of the operation scenario for IoT sensors aided beamforming system

reference to the north, *d* is the actual distance between the mobile client and the AP while the mobile client is on the move.

With the longitude and latitude of both of the AP and mobile client are known, the distance and bearing between the 2 points was calculated using the following formula, formula (7.1) is for the distance and formula (7.2) to (7.4) are for the bearing.

$$d = \mathrm{acos}(\sin Lat1 \times \sin Lat2 + \cos Lat1 \times \cos Lat2 \times \cos(Lon2 - Lon1)) \times R \tag{7.1}$$

$$x = \cos Lat1 \times \sin Lat2 - \sin Lat1 \times \cos Lat2 \times \cos(Lon2 - Lon1) \tag{7.2}$$

$$y = \sin(Lon2 - Lon1) \times \cos Lat2 \tag{7.3}$$

$$be = \mathrm{atan2}(x, y) \tag{7.4}$$

where *R* is the radius of the earth (6,371 km), *d* is the distance between 2 points in kilometers and *be* is the bearing from point 1 to point 2 in degree (0° means heading north, the positive value means clockwise from the north and the negative value represents counter clockwise from the north, *Lonn* and *Latn* are the longitude and latitude of point *n* in radians.

Knowing the heading of the mobile client and the bearing between the moving client and the AP, we can determine the direction of the beamforming beam, hence the RF beam for the mobile client and the base station can be steered towards each other. The distance between the mobile client and access points is constantly monitored, to ensure it only connects to the nearest AP, when the next AP is nearer than the current access point, the mobile client will prepare to roam over and hand over the beamforming coordination to the next nearest access point.

In the case of scenario (ii) where there is limited or no GPS signal coverage, the onboard IoT sensors will come to play. When the mobile terminal is moved away from the GPS coverage zone, it will make use of the last GPS fixed location and perform data interpolation with the aids of IoT sensors and sensors fusion that combined all the sensors data to progressively determine the current location and heading information of the mobile client, following the same approach as described earlier, the beam steering direction can be calculated. Due to the location and heading information were obtained via interpolation from the last GPS fix and sensors fusion, because of the data interpolation, the accuracy of the location and heading will be deteriorated over time, to overcome this, a BLE beacon can be installed at a fixed interval along the stretch where GPS reception is poor such as bus stops or lamp poles. Each beacon will have known coordinates, when the mobile client associated with the roadside beacon, it will be able to reposition itself and offset the accumulated location and heading error.

The fixed roadside infrastructures and fix operation route for public transportation and combining with the sensor fusion from existing IoT sensors such as GPS, accelerometer, gyroscope, odometer, Bluetooth etc., and leveraging on the operation behaviour and IOT infrastructure, we can reduce the highly complex field-programmable grid array (FPGA) cost that is usually used in the conventional beamforming smart antenna system, this is expected to reduce the smart antenna cost further while proving more accurate beamforming capability. The proposed modern smart antenna system is expected to set an important milestone in the smart antenna industry.

7.4 Conclusions

The adoption of wireless communication media in the transportation sector is an unstoppable phenomenon simply due to the fact that more and more information is being digitised, public security concern and raise of commuters expectation in term of internet connectivity and infotainment, the demand is expected to multiply with the recent government initiative in the IoT activities. Even though the spectrum regulator such as FCC and the local regulatory body continues to explore the possibility to open out more frequency spectrum for unlicensed usage, however, the air space congestion issue persists with the over helming demand in the wireless communication, therefore, the research and development in the smart antenna area will remain as an important topic at least for now until the near future when the wireless communication can be replaced by some other destructive technology that will change the wireless game entirely.

When we talk about infrastructure deployment, the most common question posed is what is the implementation cost? How long we need to wait before we can see the ROI? There is no exception to infrastructure deployment in the transportation sector. We have learned some practical technique in this book on some cost reduction measure via practical approach by taking advantage on the operation scenarios for public transportation, it might be worth opening the idea to other types of wireless applications in the transportation-related area, such as car sharing, mobile surveillance system and future autonomous vehicle and robotic. The innovation in the smart antenna deployment shall continue to emerge to achieve the ultimate cost, user-friendliness, maintainability etc. from a different angle such as usage scenarios, deployment consideration etc. instead of just look at the smart antenna itself.

Future direction will make very interesting if we consider a combination of techniques from different technology that makes the smart antenna even smarter and more reliable, in this book, we have discussed how to make use of various IoT sensors that are commercially available to combine with the sensor fusion technology which can be integrated into the beamforming algorithm to produce a more reliable and less complex beamforming system, that translates into cost-saving by reducing the system complexity, there is no doubt that the new smart antenna system can align its path and make its contribution to the smart antenna industry.

Reference

1. U-blox positioning chips and modules with dead reckoning functions, https://www.u-blox.com/en/positioning-chips-and-modules#tab-dead-reckoning. Access 29 May 2020

Printed in the United States
By Bookmasters